Exploring Creation

with Human Anatomy and Physiology

by Jeannie K. Fulbright and Brooke Ryan, M.D.

Exploring Creation with Human Anatomy and Physiology

Published by
Apologia Educational Ministries, Inc.
1106 Meridian Plaza, Suite 220
Anderson, IN 46016
www.apologia.com

Manufactured in the United States of America
Seventh Printing: October 2015

ISBN: 978-1-935495-14-7

Printed by Bang Printing

Cover photos licensed through Shutterstock
© Sebastian Kaulitzki, Elena Kalistratova, Alexander Vasilyev, Artman, HKahn

Unless otherwise stated, all Scripture quotes come from the New American Standard Bible (NASB)

Apologia's Young Explorer Series
INSTRUCTIONAL SUPPORT

Apologia's elementary science materials launch young minds on an educational journey to explore God's signature in all of creation. Our award-winning curriculum cultivates a love of learning, nurtures a spirit of exploration, and helps turn textbook lessons into real-life adventures.

TEXTBOOK

Apologia textbooks are written directly to the student in a highly readable conversational tone. Periodically asking students to stop and retell what they have just heard or read, our elementary science courses engage students as active learners while growing their ability to communicate clearly and effectively. With plenty of hands-on activities and experiments, the Young Explorer Series allow young scientists to actively participate in the scientific method.

NOTEBOOKING JOURNAL

These spiral-bound **Apologia Notebooking Journals** contain lesson plans, review questions, additional projects, full-color mini-books, puzzles, and much more to keep students not only actively engaged in learning but also organized.

JUNIOR NOTEBOOKING JOURNAL

Designed for younger students and those who struggle with writing, **Apologia Junior Notebooking Journals** cover everything in the regular notebooking journals, but at a more basic level and with primary writing lines. With simpler vocabulary pages, fewer crossword puzzles, and additional coloring pages, junior notebooking journals make science enjoyable for even your youngest student.

AUDIO BOOKS

Some students learn best when they can see and hear what they are studying. Having the full audio text of your course is great for listening while reading along in the book or riding in the car! **Apologia Audio Books** contain the complete text of the book read aloud to your student.

FIELD TRIP JOURNAL

Apologia's Field Trip Journal is a fun and exciting way to record those moments when textbook lessons turn into real-life adventures. You and your students can successfully prepare for field trips, map the places you visit, and document entire field trips from planning stage to treasured memory.

At Apologia, we believe in homeschooling. We are here to support your endeavors and to help you and your student thrive! Find out more at apologia.com.

Scientific Speculation Sheet

Name _____

Date _____

Experiment Title _____

Materials Used:

Procedure: (What you will do or what you did)

Hypothesis: (What you think will happen and why)

Results: (What actually happened)

Conclusion: (What you learned)

Table of Contents

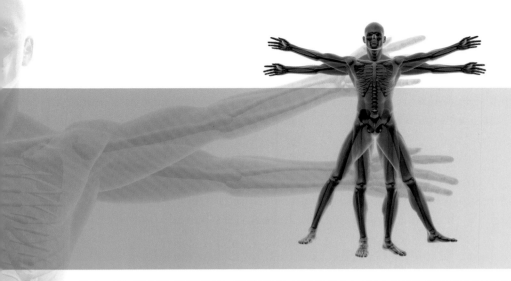

INTRODUCTION

Welcome to *Exploring Creation with Human Anatomy and Physiology*. This unique course explores the human body using engaging text, exciting activities, interesting experiments, and proven methods to help your students remember what they've learned. The text is written directly to the students, making it very appealing to kids from six to thirteen. The material is presented in a conversational style that will make science enchanting and memorable for your students, creating an environment in which learning is a joy.

Lesson Increments

There are 14 lessons in this text. Each lesson should be broken up into manageable time slots depending on your family's schedule and the ages and attention spans of your children. Most lessons can be divided into two-week segments. The *Anatomy Notebooking Journal*, designed to accompany this text, includes a suggested schedule that follows a two-week sequence. This schedule takes the "guess work" out of how to efficiently complete the course. However, if you choose to complete the course without the notebooking journal, simply read half the lesson during the first week and the other half during the second week, leaving time for the various assignments, projects and experiments.

I believe notebooking and narration are superior methods for facilitating retention and for providing documentation of your children's education. At the end of each lesson, your students will be assigned various activities that utilize these methods. Their learning will be enhanced through narration, notebooking activities, projects and experiments. In addition, your students will complete a Personal Person Project that is unique to this course. Each of these learning activities is detailed below.

Narration

The act of narrating is an ancient and effective method, stimulating a child's mind to focus his attention on the information he has just heard or read. When the student is required to "retell" this information through narration prompts, he becomes engaged with the material as an active learner. Greater retention is achieved as a result.

Throughout each lesson of the text, the student is asked to "tell back" or narrate the information learned from what he has read or heard. The "What Do You Remember?" questions near the end of each lesson can be used for oral narration. If your student has the *Anatomy Notebooking Journal*, you'll find templates for these questions that can be used for written narration. Whether oral or written, narration will propel your students forward in their ability to effectively and clearly communicate with others what they have learned.

Notebooking Activities

At the end of each lesson, various notebooking activities are assigned to the students. They include illustrating concepts, writing essays, and completing other types of written exercises. These will be placed in each student's notebook. There are templates provided for each of these activities in the *Anatomy Notebooking Journal*. You will find these notebooking activities to be important tools for providing a record of progress and learning as well as documentation for review. Notebooking is flexible and allows for multi-aged learning. A twelve-year-old student may write a brief essay, while a six-year-old may illustrate his learning through a simple drawing. Both, however, will retain what they've learned through their unique and personal expressions of the topic.

Personal Person Project

This is a fun and unique project that allows students to create a paper model of the human body. Throughout the course, students are given instructions for adding the organs they've studied to a human form they have created. This project is designed to engage the students in a hands-on application of what they've learned – serving to solidify not just the facts about the organs, but where they are actually located in the body. The human form is personalized with the student's face and is placed in his course notebook.

As students progress through the lessons, they will draw and cut out the organs, taping them to their Personal Persons. Again, if your students have the *Anatomy Notebooking Journal*, images of each organ are located in the appendix for your students to cut out.

Projects and Experiments

Every lesson ends with a project or experiment. Many of these experiments will help your children develop the skills needed to conduct valid and scientifically accurate experiments, both now and in high school. It is recommended that your students complete at least a few of these experiments so as to become familiar with the scientific method. Though many experiments utilize household items, some require science supplies only available through specific sources. A list of supplies is provided starting on page 15.

Course Website

If your students would like to learn more about the topics discussed in this course, there is a course website that allows them to dig even deeper into the topic of human anatomy and physiology. To go to the course website, simply type the following address into your web browser:

http://www.apologia.com/bookextras

You will see a box on the page. Type the following password into the box:

Godmadeyou

Make sure you capitalize the first letter, and make sure there are no spaces in the password. When you hit "enter," you will get an "access granted" message. Click on the link there to go to the course website.

How To Use This Book
A Step-By-Step Guide

1. If you have not purchased a supplies kit, you will want to scan the materials list located on pages 15-17 to see what you need for the lesson you are going to do.

2. Begin by reading the lesson to the students (older students may read the lesson themselves). There will be places during the lesson where the students are asked to "tell back" or narrate what they have learned up to that point. These are not written narrations; they are impromptu oral presentations.

3. Occasionally there will be a "Try This!" activity wherein the learners are encouraged to get a few supplies and try a little project or experiment to demonstrate a point made in that section of the book. Ideally, the project should be done right then. However, don't be discouraged if you do not have the materials. You can always go back and do the project later.

4. You will continue reading until you feel a natural break is at hand. Each family will differ in the amount of reading done in each session. Some families become extremely engrossed and will want to read an entire lesson. Most families will read a quarter to half the lesson. There are many places within each lesson that are natural stopping points. You decide when to stop reading. The book is designed to give a lot of flexibility with this, so that you can complete the book in a year in a way that works for your family.

5. When you end for the day, ask your children to orally tell you what they learned. They do not need to write anything down until they reach the end of the lesson.

6. When you reach the end of a lesson, you will come to a "What Do You Remember?" section. This is a series of specific questions to ask your children in order to prompt their memories about the lesson. Don't expect young children to remember most of these. Don't expect older children to remember all of them. However, this is a great time to enter into discussion about what they learned. These are also oral, not written.

7. After your children tell you what they remember, it's time for the notebooking activity. In this activity, each child will be asked to record in writing all that she wants to remember about the lesson. I would not force her to record every detail of the "What Do You Remember?" section. Also, do not have her write down what you want her to remember. Allow her to decide what she thought was interesting and important. Let her decide what she wants to remember. For non-writers or slow writers, you can type out or write out what they tell you. If your child is struggling to recount her learning, you can encourage her with questions. Make this an enjoyable experience without a lot of correction and nit-picking. Eventually, your child will be able to accurately and systematically recount what she learned. Many children graduate from high school never learning this skill.

8. Occasionally, the notebooking activity will also include some sort of work beyond just recording the information the students found interesting or want to remember. They might be asked to diagram something or produce a creative work associated with the subject.

9. If it is mentioned in the lesson, have the students add an organ to their Personal Persons.

10. The last thing students should do is the experiment for the lesson.

Items Needed To Complete Each Lesson

Every child will need his own notebook (or the *Anatomy Notebooking Journal*), blank paper, lined paper, and colored pencils.

Lesson 1
- One apple
- Apple peeler
- Eight cups
- Table salt
- Epsom Salt
- Baking Soda
- Kitchen Scale
- Measuring cups
- A piece of clear plastic (Plastic wrap works fine.)
- A medicine dropper
- Water
- A sheet of 8-inch x 10-inch paper (Flesh-colored construction paper is recommended.)
- A pencil
- Scissors
- A photograph of your face (between 2 and 3 inches tall)
- Tape
- A sharp steak knife and a parent to use it
- A spoon
- A plate
- A glass or ceramic cereal bowl
- Cooking spray, like Pam
- A box of yellow colored Jell-O
- A box of unflavored Knox Gelatin
- A jelly bean or a peanut M&M candy
- Several Skittles, Everlasting Gobstoppers or M&M candies
- A Starburst Gummiburst or several Smarties
- A Fruit Roll Up
- Nerds or cake sprinkles
- Tubular cake sprinkles or Twizzler Pull and Peels
- A large gumdrop, jaw breaker, or round chocolate truffle

Lesson 2
- Modeling clay
- Toothpicks
- Two eggs
- A plastic container with a lid that seals tightly (only slightly bigger than the eggs)
- Water
- Masking tape
- A tape measure
- A cooked chicken wing
- A pair of rubber or plastic gloves
- White vinegar
- Two plastic containers with lids (just big enough for a chicken wing and some liquid)
- Plastic wrap
- A parent with a knife

Lesson 3
- A calculator
- Bathroom scales
- Beef brisket
- Gloves
- A toothpick
- A magnifying glass
- A timer
- A nylon stocking that you can destroy
- A ball of clay
- Scissors
- A clothes pin that opens when you squeeze on it and closes when you release
- Graph paper (Only older students need this.)
- A timer
- A pencil
- Paper

Lesson 4
- An old baby tooth (An animal tooth will do.)
- A soda pop like Coke or Pepsi
- A saltine cracker
- A mirror
- Cheese
- Two Ziploc bags
- A piece of bread
- Water
- A measuring tape
- Cooking oil
- A bowl
- Water

Lesson 5
- A cool window
- Iodine solution (available at drug stores)
- Several items of food to test
- A brown paper bag
- Scissors
- A hair dryer
- Food from the pantry
- A banana
- A vitamin C tablet
- Juice, freshly squeezed from different fresh fruits or vegetables that you think might contain vitamin C (oranges, tomatoes, strawberries, peaches, etc.)
- Cornstarch
- A medicine dropper
- A juice glass
- Several small cups or test tubes
- A measuring cup
- Measuring spoons
- A small pot
- A stove
- A spoon
- A Scientific Speculation Sheet

Lesson 6
- Honey
- Two pieces of cardboard
- A mirror
- Plastic food storage container
- Thin and thick rubber bands
- A grape
- A straw
- Cellophane tape
- An empty plastic large-mouth drink bottle
- Scissors
- Two balloons
- Tape
- A 2-liter plastic soda bottle
- A 1-foot-long piece of flexible tubing (like the kind you use for aquariums)
- A mixing bowl
- A measuring cup
- A jump rope
- A timer

Lesson 7
- A mirror
- A flashlight
- A bowl
- 1 cup of corn syrup
- ¾ cup of candy red hots
- A white jelly bean
- Candy sprinkles
- Iron-fortified cereal
- A Ziploc bag
- A strong magnet
- A mallet
- A blood typing kit

Lesson 8
- Graham crackers
- Blue frosting and red frosting (or white frosting that has been colored blue and red with food coloring)
- Toothpicks
- Large and small marshmallows
- A toothpick
- A small ball of clay
- A nine-inch balloon
- A small plastic funnel
- 18 inches of vinyl tubing (from a hardware store)
- Tape

Lesson 9
- Six different colors of clay
- Different colored pieces of paper on which you will write down your questions. Each color will represent a different body system.
- A file folder to create your game board
- Colored markers to draw your game board
- Game pieces made out of self-hardening clay in the shape of people or body parts
- Dice

Lesson 10
- A scrap of paper about 4 inches square
- Six different colors of clay
- Four pennies
- A ruler
- Someone to help you
- Two eggs
- A plastic Easter egg (larger than the real eggs)
- Karo syrup or molasses

Lesson 11
- A few bites of something you like to eat
- A bottle of vanilla
- A variety of herbs from your kitchen
- Four paper plates
- A pencil
- A mirror
- Five small custard cups
- Five Q-tips
- Saltwater
- A lemon
- Sugar
- Unsweetened cocoa or ground coffee
- Saltine crackers
- A glass
- A mug
- Chocolate milk
- Foods with familiar tastes and textures
- Something with which you can cut the foods
- A Slinky
- A blindfold
- Someone to help you
- A pencil
- A cup
- A timer
- A darkened room with a mirror
- A flashlight
- Colored pencils or crayons
- A few friends
- A piece of paper for each friend
- A magnifying glass
- A sheet of paper
- A pencil
- Index cards
- A partner
- Two markers with brightly colored lids
- A Nerf ball

- A volunteer willing to taste foods while blindfolded
- A variety of foods with sweet, salty, bitter, sour, and umami (savory) tastes
- Straws for testing liquids
- Spoons for putting the food on the volunteer's tongue

Lesson 12
- A damp cloth
- A mirror
- A strand of hair
- Three bowls of water: one hot, one cold, one lukewarm
- A rubber band
- A stamp pad
- Several sheets of paper
- A photocopy of the braille chart
- White school glue
- A volunteer
- Five large paper clips
- A ruler or tape measure
- A chart with all the body parts you intend to test

Lesson 13
- A bacteria testing kit with agar and Petri dishes
- Q-tips
- Tape

Lesson 14
- Photos of when you were a baby, toddler and small child

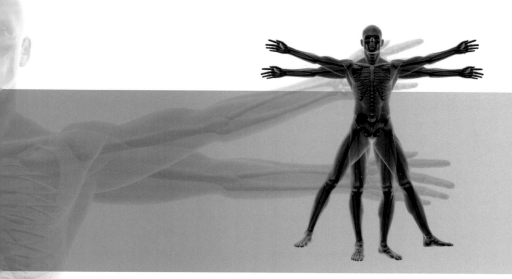

INTRODUCTION TO HUMAN ANATOMY AND PHYSIOLOGY

Have you ever wondered what happens to the food you eat, or why your mom tells you to eat your vegetables? Have you considered why your body forms scabs over certain injuries? Are you curious about your brain, your heart, or your bones? Well, even if you haven't thought of these questions, I think you'll enjoy this book about human anatomy and physiology. It will answer all these questions and many, many more.

Have you heard the words **anatomy** (uh nat' u mee) and **physiology** (fiz' ee awl' uh jee) before? Most people have some idea what the word anatomy means, but they aren't quite sure about physiology. Is that true for you? Well, anatomy and physiology have two different

Studying anatomy and physiology allows us to understand what our bodies are made of and how they work.

meanings, but they're really quite similar. Anatomy is the study of the human body, all its parts, and how it's put together. Physiology is the study of how all those parts work. For example, anatomy looks at how your heart and lungs are designed, while physiology studies how they work together to carry oxygen to your entire body.

Studying the human body is important because you live in your body. You need to know all about how it works, but remember, you are actually more than just a human body! You see, God made lots of creatures, but when He made people, He did something different. *He made humans in His image.* What does that mean? Well, the Bible says that God is spirit. And those who worship Him must worship in spirit and truth (John 4:24). So, when He made man, He gave us a spirit that was created in His image. The spirit that God gave you allows you to have a relationship with Him. It gives you the ability to know God and communicate with Him. Your spirit

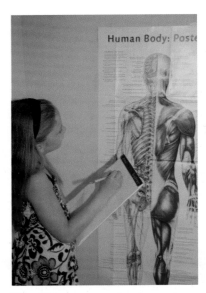

We will learn more about ourselves and God as we study how He made us.

makes you unique from all the other creatures God made. Isn't that wonderful? So the next time someone says we are just animals, you'll know the reason that's not true.

The Bible tells us that God specially chose each one of us to be born and that we were carefully put together by Him. We didn't come into this world by accident or as a mistake. Psalm 139:13-14 tells us, "*For you formed my inward parts; you wove me in my mother's womb. I will give thanks to You, for I am fearfully and wonderfully made; wonderful are Your works, and my soul knows it very well.*" Do you understand what that means? All of God's works are wonderful. You are one of God's works. That means *you are wonderful!*

God made your body in such a marvelous and complex way that scientists are always, even today, discovering new things about how your body works. People have been interested in discovering more about this marvelous miracle, the human body, since the beginning of time. Soon after their creation, Adam and Eve were likely fascinated by their beautifully designed and incredibly useful bodies. Perhaps they were amazed at the strength they had or how long they could run. Perhaps they wondered why they felt better when they ate certain foods, but not as good when they ate others. Indeed, ancient writings tell us that people have been studying the human body since the earliest recordings of history.

History of Anatomy and Physiology

Before we get into the nuts and bolts of this course, let's take a quick look at what people once believed about the human body and how we became much more knowledgeable about it over time.

Ancient Egyptians

Perhaps you've already studied the ancient Egyptians. If so, what are some things you remember about them? What did they do with people when they died? Even as far back as 3400 BC (before Christ), the Egyptians preserved dead bodies. They used salt and other chemicals to keep them from rotting and decaying. This process dried the bodies, making them into mummies (mum' eez). The Egyptians did this because they believed that a person needed his or her body to get to heaven.

In order to mummify the body, they first dissected (or cut apart) the body and removed some of the organs: the brain, liver, stomach, and intestines. They left the heart because it was thought that the heart was needed in the next life. The Egyptians' religion taught that a person's entrance into heaven was based upon the weight of his heart. They believed that when a person died, the gods would weigh his heart, and if it were heavier than a special feather one of the gods kept, the person wouldn't get into heaven.

Even though the Egyptians had religious beliefs that did not recognize God as their Creator, their ability to dissect and preserve bodies shows how far ahead they were in their understanding of anatomy. In fact, the Egyptians' writings reveal they could do simple surgeries and even fix broken bones! That's pretty amazing.

The ancient Egyptians knew a lot about the human body and were able to preserve it by mummification.

However, because they did not understand the true God, they could not get very far in their understanding. For example, they believed that magic performed by their gods made people sick and made the body function as it did.

You can use the same basic chemicals that the Egyptians used to mummify a piece of fruit. Let's experiment!

Try This! You will need an apple sliced into 8 equal slices, table salt, Epsom salt, baking soda and 8 cups for each apple slice. Weigh each slice with a kitchen scale. Put the slice into a cup and record the weight on the cup. Pour the following ingredients into each cup and label the cup accordingly: Cup 1: ½ cup baking soda. Cup 2: ½ C. Epsom salt. Cup 3: ½ C. table salt. Cup 4: ¼ C. Epsom salt & ¼ C. table salt. Cup 5: ½ C. table salt & ½ C. baking soda. Cup 6: ½ C. baking soda & ½ C. Epsom salt. Cup 7: 1/3 C. baking soda & 1/3 C. Epsom salt & 1/3 C. table salt. Cup 8: control apple with no ingredients. Let cups sit for 7 days. Then remove, dust off, and weigh each apple slice. Which mixture removed the most moisture and preserved the apple the best?

Ancient Hebrews

If you've read the Bible, you've probably heard of God's chosen people, the Israelites. In the Bible, they are also called the Hebrews. Though they didn't study the human body like the Egyptians, the Hebrew people had the benefit of good health when they kept God's laws. You see, God gave the Hebrews a lot of rules; some of those rules had the advantage of protecting the Israelites from tiny organisms (or' guh niz' uhmz – living things) called germs. These rules can be found in the Old Testament. This part of the Bible tells the story of God's people before Jesus came to earth. Thousands of years before scientists knew that germs existed, God's rules helped the Israelites keep them away!

Creation Confirmation

By examining some of the rules that God gave the Israelites, we have even more evidence that our God is real. Let's take a look at an example of one rule.

In the passage below (Leviticus 13: 3-4) God has given the Israelites specific rules for how to handle someone with a rash. Let's read it together:

The Israelites were given God's laws in the Old Testament, which they kept on scrolls in beautiful cases.

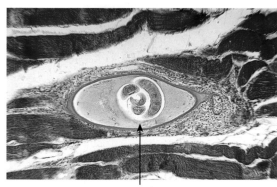

parasite

This microscopic photograph shows you muscle tissue infected with a parasite that is common in pigs.

The priest is to examine the sore on his skin, and if the hair in the sore has turned white and the sore appears to be more than skin deep it is an infectious skin disease. When the priest examines him, he shall pronounce him unclean. If the spot on his skin is white but does not appear to be more than skin deep and the hair in it has not turned white, the priest is to put the infected person in isolation for seven days.

We see here that God knew that rashes are contagious (meaning other people can catch them). In fact, the instructions to isolate the man with a serious rash are called a quarantine (kwar en teen). Even today, we quarantine people by keeping them in one place, separated from others, so they cannot spread disease. The Israelites may not have known about germs and how disease was spread, but God sure did!

There are many laws that God gave in the Old Testament that we can now see are helpful for keeping ourselves safe, healthy and clean. Some of God's laws included instructions for washing the body and clothing. Today, we know that germs can live on the body and in clothing.

Does your mom tell you to wash your hands before you eat? Wise mom! Who knows what germs you may have come in contact with during the day! Maybe you touched a doorknob that someone with the flu touched. Perhaps you picked up a nickel in the dirt that had pinworm eggs on it, or high-fived your best friend who, unknowingly, was contagious with strep throat. It's important that you wash your hands to keep from getting sick. Would you believe that there are people on this earth who rarely (if ever) wash their hands? It's true, especially in places where people are not educated. Yet, thousands of years ago, the Israelites were washing regularly. When people met with the Israelites, they considered them extremely strange for all the washing they did!

This is great evidence for those who do not believe in God. If God did not exist, how is it possible that the instructions given to the Israelites of the Old Testament are scientifically accurate and beneficial? The answer is: God is the Original Scientist of the world. Although we know today that one of the benefits from following the many rules that God gave the Israelites protected them from getting sick, God's reason for giving the Israelites these specific laws involved much more than we could talk about here.

You will remember what you have just learned a little better if you explain it to someone else.
Throughout the course, you will do this often.
Take time now to explain all that you've learned so far.

Ancient Greeks

Long after God gave His laws to the Israelites, a group of people along the Mediterranean Sea were pondering grand things in their minds. These men were called the ancient Greek philosophers. Around 500 years before Jesus was born, they were fascinated by the natural world. After close inspection of plants and animals, they carefully wrote down what they learned. Although the Greeks, like the Egyptians, believed in many gods, they also believed that by using their brains, they could understand more about the world around them. By doing this, they ended many of the superstitious beliefs that the Egyptians had passed on about the human body.

As part of their discovery process, the Greek philosophers would come up with a **hypothesis** (hi pah' thuh sus – an educated guess) that they hoped would explain something they saw. After a lot of discussion, if the hypothesis seemed correct to everyone, they would claim it as a **theory** (theer' ee). A theory in ancient Greece made something a belief. These beliefs were developed after lots and lots of debates and discussions.

Unfortunately, this way of deciding a belief left out a very important step in scientific studies – experimentation. They preferred to find answers by thinking about and discussing their ideas. Although these ancient philosophers produced valuable knowledge with their discussion, sometimes that knowledge could have been expanded and understood better with the use of scientific experimentation. Experiments are designed to test hypotheses to see if they hold true.

Imagine what happens when people don't experiment to test their ideas. Do you think the Greek philosophers developed a lot of really good theories this way? As you may have guessed, many of their theories weren't correct. For example, a very famous Greek Philosopher named **Hippocrates** (hi pah' kruh teez) believed that in order to be healthy, your body needed equal amounts of four liquids. They called these liquids **humors** (hyoo' murz), and they listed them as black bile, yellow bile, **phlegm** (flem), and blood. If you got sick, they thought it was because you had too much of one of these humors. So, they tried to cure you by removing some of the humor.

This statue of Hippocrates is in Larissa, Greece, where he died.

This ancient manuscript records the Hippocratic oath in the form of a cross.

If they thought you had too much blood, they would remove some blood from you – usually by applying leeches to your body.

The Greeks even believed that a person's personality was affected by his humors. If someone was very calm and relaxed, they thought he had a lot of phlegm, which made him more patient. If he was very energetic and full of ideas, they thought he had extra blood. If he was very controlling and always in charge, they thought he was full of yellow bile. If he was a deep thinker and very emotional, they thought he was full of black bile.

Sadly, the theory of the four humors was believed all the way from Hippocrates's time until the 1800s, when most historians think modern medicine was developed. In fact, Hippocrates is considered the Father of Modern Medicine, even though he was mistaken about the four humors. This is because he was the first major philosopher to reject the idea that supernatural forces caused illnesses. In addition, doctors today are required to take an oath created by Hippocrates, himself. This Hippocratic oath states that doctors should always do good and never harm to anyone.

Try This!

Many personality tests still use the four humors to describe different types of people, even though we know that these "humors" don't determine personality. Here is how they are broken down:

Phlegmatic (fleg mat' ik – phlegm): Easy going and easy to get along with, usually happy
Sanguine (sang' gwin – blood): Excited, full of energy, usually happy
Choleric (kol' uh rik – yellow bile): Controlling, leader, likes to be in charge
Melancholy (mel' uhn kol' ee – black bile): Thoughtful, artistic, emotional

Now, think of the people in your family or friends you know. Consider which one of the personalities they have. Write it down and see if others agree with you.

Aristotle

Aristotle (ar' ih stot' uhl), another famous Greek philosopher who lived about 100 years after Hippocrates, was one of the first true scientists and one of the greatest thinkers of all time. He believed that you needed to conduct experiments, not just discuss your thoughts. He dissected plants and animals to learn about their inner parts, and some think he even dissected dead people, although not all historians agree about that. He helped others understand how important it was to examine and test everything they believed. Unfortunately, because Aristotle was not taught about the one true God, many of his theories were not correct.

One of Aristotle's theories we now know to be wrong was **spontaneous** (spon tay' nee us) **generation** (jen' uh ray shun). This theory said that life could come into existence from nonliving things. You see, Aristotle noticed that within a few days of leaving raw meat out, maggots suddenly appeared on the meat. He thought that living things could suddenly spring forth from nonliving things, like maggots springing forth from meat. He saw it happen so often, he was certain it was true. Of course we now know that maggots come from eggs laid by flies that land on the meat. The eggs were so small that Aristotle couldn't see them. Today we know that life does not come from things that are not living. Just as the Bible says, life only comes from life; maggots come from flies, not dead meat. Yet because Aristotle was such a great thinker, his idea of spontaneous generation was accepted as true for hundreds of years.

Creation Confirmation

In case you have not yet heard, there is a theory (or belief) floating around today that says all life sprang into existence spontaneously. Does that sound familiar? Yes, even today there are people who still

This famous painting by the Italian artist Raphael has Plato (left) and Aristotle (right) as its central figures. It is called "The School of Athens."

hold on to Aristotle's theory. Some people believe that the entire world and everything in it came together accidentally from nonliving things that were floating around in space. They believe it happened without God, without a plan, and they believe that some life forms changed (or evolved) into different forms of life to become all that we see today. As you can see, it's obvious that some beliefs – like Aristotle's – are hard to remove from modern science.

In your own words, explain what you have learned about the history of anatomy and physiology. Include information about the Egyptians, Hebrews, and Greeks.

Ancient Rome

About 200 years after our Lord Jesus was born, a brilliant Roman physician named Galen (gay' lun) arrived on the scene, moving the study of anatomy along. Galen read all of Aristotle's works and followed much of his teachings. He had lots of opportunities to test his theories on actual people, because he was a doctor for Roman gladiators who were always getting hurt. People were stunned when the beaten, bloodied gladiators survived after being treated by Galen. One of Galen's secrets was to always use clean rags when treating his patients. Before this, doctors would use filthy rags filled with germs (that no one knew existed) to clean the wounds of

injured people. Galen also soaked the rags in wine. He didn't know why this made his patients better, but it did. Of course, now we understand today why it worked. Wine contains alcohol, which kills certain germs!

Since Galen was much more successful at treating patients than most other doctors of his time, it was not long before the Emperor wanted Galen as his own personal doctor. To help himself understand anatomy, Galen dissected animals, especially monkeys. He never dissected a person, because it was against the law in Rome to do that. Galen wrote a great deal about his work and made many illustrations (scientists like to draw what they learn). These writings and illustrations helped other scientists better understand anatomy for a thousand years.

This is what a 19th century artist thought Galen may have looked like.

Creation Confirmation

Jesus, because He is God, knew a lot more about health than the people that lived during the New Testament times. For example, He knew the disinfectant properties of wine. We can see this in the parable of the Good Samaritan. As a part of the parable, Jesus said the Good Samaritan, "*came to him and bandaged up his wounds, pouring oil and wine on them*" (Luke 10:34). This is additional evidence that Jesus was God, as He had knowledge of the healing properties of alcohol even before it was known by others.

European Scientists

Much later, in the 16th century, a young French scientist named **Andreas** (ahn dray' us) **Vesalius** (vuh say' lee us) began to question Galen's ideas. Vesalius was an anatomy professor at the age of 23! What was your dad doing when he was 23? That's pretty young to be teaching anatomy at a university. Vesalius dissected human **cadavers** (kuh dav' urz – dead bodies), proving that many of Galen's teachings were wrong. He published a great big series of books on his findings. Remember, Galen's ideas were based on studying the insides of animals. By looking at the insides of people, Vesalius could correct some of Galen's mistakes. Many scientists were devoted to Galen's teachings, and they were angry that Vesalius was proving Galen wrong. These men made life difficult for Vesalius, so he left his teaching job and worked as a doctor for Emperors.

But it wasn't long before more and more people became interested in studying anatomy. Soon, the world was teeming with scientists. There were many people who began to discover how the human body works. We call this time in history the Renaissance period. You'll learn all about it in your study of history.

One very important discovery was made by a Dutchman named **Anton van Leeuwenhoek** (lay' uh wun hook'). He discovered a great use for glass lenses, which opened up a whole new world for scientists. He found that if he made a lens correctly, he could use it to magnify things. He was such a good lens maker that he could magnify things to nearly 200 times their natural size! In other words, this Dutchman built microscopes! Fascinated with what he saw, he studied everything he could get his hands on and made lots of detailed drawings of his findings.

What was van Leeuwenhoek able to see through his microscope? Well, one thing he saw was red blood cells flowing single file through a blood vessel in the tail of a tadpole. He also saw many microscopic organisms that he found in pond water and even in the saliva taken from a person's mouth! He called these organisms "animalcules," because he likened them to little animals. His invention enabled him to see things no human eye had ever seen before! Without the invention of the microscope, science would never have come as far as it has today!

Robert Hooke was a scientist who was interested in the same, tiny world that van Leeuwenhoek found interesting. He made a microscope that is much like the microscopes we use today. It is called a **compound microscope**, and it uses two lenses to magnify objects. If you position the lenses correctly, you can actually magnify things more than you can when you have only one lens. One object Hooke looked at under his microscope was a piece of cork. He saw that the cork was actually made up of many rectangular blocks. They looked like the cells (rooms) that monks slept in, so he called them **cells**. We still call them cells today! What Hooke saw were actually cell walls, which in plants (cork comes from a tree) are thick and sturdy. Hooke's discovery soon led to the concept of the cell as the basic building block of life. In fact, that's what cells are, the basic building blocks of life.

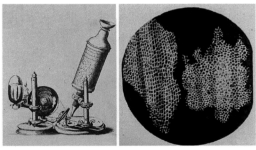

These are drawings Robert Hooke made of his microscope (left) and the cells he saw with it (right).

Try This!

JESUS

You may already have experience using a magnifying glass, but let's do an experiment with water to show how a magnifying glass works. You will need a piece of clear plastic. You will also need a medicine dropper. Put your piece of plastic over the word inside the white box to your left. Now put a single drop of water on the plastic, right above the word. Can you read the word through the water drop? You probably can, because the water drop bends light so as to magnify the things underneath it. Experiment with different sizes of water droplets. Which ones make the word the biggest?

Explain to someone all that you've learned today before moving on to the next section.

Cells

Because scientists like van Leeuwenhoek, Hooke, and many who came after them could not see the tiny, inner parts of a cell, most scientists thought that cells were very simple things. They thought each cell was just a block with a little dot inside, which they called the **nucleus** (new' klee us). The word nucleus means core or center. Today, with more advanced microscopes as well as other advanced laboratory equipment, we are beginning to learn just how complex cells really are. Rather than being an uncomplicated block, each one is like a miniature city! These "cities" are so well run and so highly structured that what we once believed was something very simple and easy to understand, we now know is still beyond our complete understanding!

Believe it or not, this drawing represents a simplified view of the kind of cells that are in your body.

Cell Anatomy

Let's study the anatomy of a cell. I would like for you to get a piece of paper and illustrate each part of the cell that we discuss. Your illustration will not be as fancy as the one given above, but it will still be better than anything Robert Hooke or Anton van Leeuwenhoek had! After you illustrate something, draw a line out to the side and label it. If you like to write, you can also write about what it does. Later on, you will learn about the notebook you are supposed to be keeping throughout this entire course. You can use the *Anatomy Notebooking Journal* that was made for this course, or you can get a blank notebook that you will write and draw in. If you are using the *Anatomy Notebooking Journal*, make your drawing on the page that says "Cell Anatomy." If not, make your drawing on one of the first pages in your blank notebook.

Cell Membrane

In the body, cells come in all different shapes and sizes. We'll draw a cell that is oval, because it's a bit easier to draw. So, draw an oval with a boundary around it, like you see in the drawing on the next page. That boundary is the cell membrane.

Do you remember the story of Jericho found in the Bible (Joshua 6)? If so, you'll remember that the city of Jericho had a wall around it. This wall protected Jericho from foreigners and danger. It allowed certain people and things in, like friendly neighbors and food. It also kept certain things out, like unfriendly neighbors and their weapons. Well, back in those days, almost every city had a wall around it. At the wall were gatekeepers.

These were men whose job was to open and close the gate, depending on who wanted in and who wanted out.

Like a city wall, the cell membrane is the first level of protection for the cell. It works like a gatekeeper, allowing certain substances to pass in and out. Usually, only those substances that are needed (like oxygen, water, and nutrients) are allowed in. Sometimes other stuff sneaks in, but we'll discuss that in the lessons on the lymphatic (lim fat' ik) and immune systems. In addition to letting certain things in, the cell membrane also lets certain things out. The things it allows to leave the cell might be waste products or chemicals made by the cell to help your body. As you can see, the cell membrane is very important.

Inside the cell membrane are a variety of smaller structures called **organelles** (or' guh nelz). Organelle means little organ, and like an organ, each organelle has a special job. As you learn about each organelle, you will add it to your cell drawing. The organelles all float in a substance called **cytoplasm** (sye' toh plaz' uhm). Cytoplasm feels like runny jelly, but it's mostly water. However, cytoplasm also has nutrients in it. Together the organelles and the cytoplasm fill the cell.

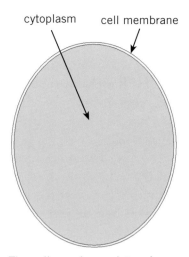

cytoplasm   cell membrane

The cell membrane determines what can enter and leave the cell. Inside, there is a jelly-like substance called cytoplasm.

Mighty Mitochondria

What if you wake up one morning and turn on the light switch, only to find it doesn't work? You walk around and realize that nothing works: your computer, your oven, your toaster. Nothing turns on! Later you realize the problem is not just with your house, but with every house on the street; soon you realize it's every house in the city. What happened? It seems that somehow the city's power supply was cut off. The power plants that supply power to your city are not working.

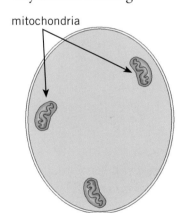

mitochondria

Mitochondria power the cell by burning fuel that they get from the food you eat.

Just as your city gets its power from power plants, each cell has power plants. Actually, it usually has many. These power plants are called **mitochondria** (my' tuh kon' dree uh). They are mighty mitochondria because they give each cell its might and power. Mitochondria are shaped like beans that have a squiggly line through the center. Take a minute to draw a few mitochondria anywhere inside your cell illustration.

How do mitochondria actually work? Well, your body digests the food you eat, turning it into smaller nutrients that go to the cell. The cell breaks some of those nutrients down into fuel (kind of like gasoline, the fuel that makes your car go). The fuel is then sent to the mitochondria, where it is actually burned (a bit like the way a car burns its gasoline). Burning the fuel makes energy – the very energy you use when you walk, run, and even think! That's pretty mighty, isn't it? Now don't think there are little fires going on in all the cells of your body. Your cells burn their fuel very gently, but they do, indeed, burn it. So, what exactly is the fuel that your cells burn? Well, mostly it's a kind of sugar called **glucose** (gloo' kohs). We'll learn all about that in the lesson on nutrition.

Your body is made up of cells – many different kinds of cells. Some cells are muscle cells, some are brain cells, some are skin cells. There are cells everywhere in your body, and all of them need energy. However, certain cells need more energy than others. Many of your muscles are always working, even when you sleep. So, your muscle cells need lots of energy. Amazingly, God created muscle cells with more mitochondria than other cells (like your skin cells) that don't use a lot of energy.

Lysosome Patrols

Do you remember how we compared a cell to a city? We said that the cell membrane is like the city wall, and the mitochondria are like power plants for the city. Let's stick to the city analogy as we learn about **lysosomes** (lye' suh sohmz'). What are the lysosomes in the city? Well, they are the policemen. They protect the cells by destroying invaders, like bacteria. They also break apart tired, worn out old organelles and send them out of the city through the city wall.

Like police in a city, lysosomes keep the cell safe.

The lysosomes have another extremely important job: they break down chemicals that the cell membrane takes inside. This is what happens: you eat a big meal of meat and potatoes. The meat has **proteins** (proh' teenz) in it, and the potatoes have **carbohydrates** (kar' boh hye' draytz). Proteins and carbohydrates are similar to beaded necklaces in that they are made up of small units linked together. In your intestines, the "necklaces" are taken apart into much smaller units – often into individual beads.

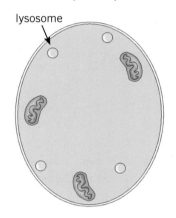

lysosome

These smaller units are taken into your cells. But, in order for your cells to use them, they often need to be broken down even further so that the cell can then rebuild them to meet a specific need. This process is very much like taking apart a Lego creation so that you can make something new. When you are taking the Lego creation apart – breaking it down into smaller Lego units – you are acting like a lysosome. So, whether they are dealing with invaders that are trying to harm the cell, worn out organelles and chemicals that should no longer be in the cell, or nutrients that the cell needs, the lysosomes break down things. This allows the cell to get rid of things it doesn't need and build the things it does need.

Lysosomes protect the cell by destroying invaders and getting rid of broken-down organelles. They also break down nutrients into smaller, usable units.

Place the lysosomes into your cell illustration. Draw the lysosomes as small round balls. Each cell has hundreds of them, but you only need to draw a few.

Grocer Golgi

What would a city be without grocery stores? Well, the cell is equipped with grocery stores called the **Golgi** (gohl' jee) **bodies**. Each Golgi body looks like a stack of differently-sized, curved pancakes. It keeps a supply of chemicals like proteins and fats that the cell makes. One thing that's unique about the Golgi body is that it sends its products for delivery, whereas most grocery stores do not. Whenever protein or fat is needed in another part of the cell city, the Golgi body wraps it up and sends it to where it is needed. That's pretty handy.

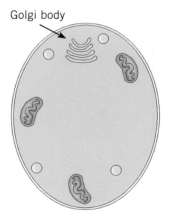

Golgi body

Draw your Golgi body as curved pancakes stacked on top of each other. There are usually several in each cell, but you can just draw one if you wish.

You've learned a lot about cell anatomy.
Try to tell someone in your own words what the cell membrane, the mitochondria, the lysosomes, and the Golgi bodies do.

The Golgi body stores chemicals and sends them to where they are needed.

ER Delivery and Pick Up

My children can't wait for the mail to arrive each day. Sometimes a big truck will bring a box with interesting new items from Grandma or Aunt June. Other times, a different truck will deliver boxes containing things we ordered. These delivery trucks are a lot like the **endoplasmic** (en' doh plaz' mik) **reticulum** (rih tik' yuh luhm). Can you say endoplasmic reticulum three times fast? Well, since no one really wants to have to say endoplasmic reticulum too many times, everyone calls the endoplasmic reticulum the ER.

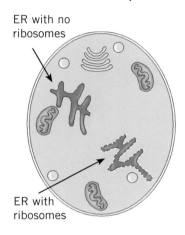

ER with no ribosomes

ER with ribosomes

The ER delivers chemicals to where they are needed and sends waste out of the cell.

The ER transports things around the cell city. It also has another job: garbage man. What would happen to your city if there was no trash pick up, or no city dump in which to discard the trash? Just imagine that. There's a great poem by Shel Silverstein about a girl who would not take out the trash, and it eventually polluted the entire city. Well, just as a city needs mailmen and garbage men, every cell needs a delivery and clean-up service. So, the ER takes chemicals that were packaged by the Golgi bodies (as well as chemicals from other parts of the cell) and delivers them to where they are needed. It also sends waste to where it needs to go so it can leave the cell. The ER is the mailman or the garbage man, depending on what needs to be done.

Just like real cities, most cells have a lot more mailmen and garbage men than grocery stores. Each cell will also have more ER than Golgi bodies. The ER resemble ribbons, some of them intermittently studded with beads. These beads are called **ribosomes** (rye' buh sohmz'), and they are very important. You'll learn about them in a moment. Draw your ER as ribbons that are connected to other ribbons. Put small beads on some of them to represent the ribosomes.

Centrioles: Mothers of the City

Cells reproduce, or copy themselves. That's one of the reasons you can grow up and become an adult. An adult has more cells than a child. As you grow up, your body makes more cells so you can get bigger and stronger. Also, sometimes you hurt yourself. For example, you might have an accident where a bone is broken. When that happens, cells die. Your body must repair the damage, and part of the repair job involves replacing the cells that died with new cells.

Centrioles (sen' tree ohlz) are special organelles that help cells reproduce. They look like two pieces of pasta that form an "L" shape. These centrioles would be the mothers of the cell city, since they help the cell make more of itself.

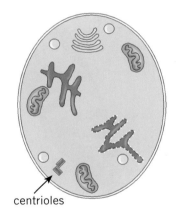

centrioles

Centrioles help cells make more cells.

The Nucleus Government

How do your parents know how fast to drive or how much tax to pay? Who decides when to pave the city streets or who will take the garbage to the landfill? Who helps those who do not have enough of what they need? All these things are decided by your city's government. Well, you may wonder, "How does each cell know what to do and when to do it?" Just like a city has a government, the cell has a control center. It is called the nucleus, and it functions like the government of the city.

The nucleus is larger than any other organelle in the cell. It's usually a big round ball and is sometimes located in the center of the cell. On the outside of the nucleus is a thick membrane called the nuclear membrane. It gives the nucleus its form. The nuclear membrane is porous, which means it has pores (holes) in it to let substances in and out. What's inside the nucleus? DNA and RNA can be found inside, as well as a smaller ball called the **nucleolus** (new klee' oh lus). The nucleolus is what makes the ribosomes that are sometimes found on the ER. You will learn about DNA and RNA in a moment.

Now is a good time to draw a nucleus for your cell illustration. You can put it anywhere inside your cell. You have completed your drawing of a cell. Now please understand that there is a lot more to a cell than what you have drawn here. Remember the drawing on page 26, and then remember that it is actually a simplified drawing! However, what you have is a great first step in understanding the main features of a cell.

Inside the Nucleus

Now that you have learned about the basics of cell anatomy, let's look inside the cell's government building, the nucleus. Learning about what's in the nucleus will help us understand more about how cells work.

The nucleus is the control center of the cell. It is surrounded by a porous membrane and contains DNA, RNA and the nucleolus.

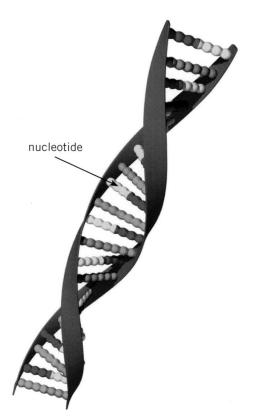

nucleotide

DNA contains genes, which are made of nucleotides.

DNA

DNA, which is short for **deoxyribonucleic** (dee ahk' see rye boh noo klay' ik) acid, is a molecule inside every person (and every living thing, for that matter). It contains all the information about that person. Your DNA is like a huge book all about you. Of course, it's longer than any book ever written! If you took the DNA out of a cell and unfolded it, it would be about 2 meters (about 6 feet) long. If you extracted all the DNA from every cell in your body and laid it out end to end, it would reach to the sun and back more than seventy times! God sure is a genius to get all that DNA packed inside your body!

The words in your DNA "book" are called **genes** (jeenz). Have you ever heard that word? You may have thought it referred to a pair of pants. Those jeans are for the outside of your body; the ones we're talking about now are on the inside. You get your genes from your parents – one set from your mother and one set from your father.

These gene "words" are made up of letters. Scientists call these letters nucleotides (noo' klee uh tides). Letters strung together make words, and words strung together make a book. Similarly, nucleotides strung together make genes, and genes strung together make DNA. Just like the information in an encyclopedia is packaged into different books, DNA is packaged into units called chromosomes (kroh' muh sohmz). Each of your cells has 46 of these packages of DNA.

What is so important about DNA? Well, the genes in DNA produce the code for a person's life. They help to determine what color hair the person will have, the color of the person's eyes, whether the person is a he or a she, when the person will lose his or her first tooth, how tall the person will be, and so on. So, a person's DNA is a *big part* of what makes the person who he or she is. The DNA is literally the recipe, or set of instructions, that helps to make the person.

Nearly every cell in your body has DNA. So, you could take a cell from almost any place in a person's body and get DNA from it. If you understood the code, you could determine a lot of information about

Scientists can use DNA to help solve crimes.

that person. No two people in the whole world have the same DNA, except identical twins or triplets. Even though identical twins have the same DNA, they are still different people, because DNA doesn't determine *everything* about a person. A person's life experiences help to determine things as well. For example, even though they have the same DNA, identical twins have different fingerprints, because they have different experiences, even as they develop before birth.

Because a person's DNA is unique (unless the person has an identical twin), detectives can solve crimes by collecting DNA at crime scenes. Everyone's skin, hair, and blood have DNA. Scientists can collect samples of skin, hair, or blood and study them to find out who was present during a crime. Determining a person's identity from DNA found at a crime scene is a rather newly-developed technique, but it is helping to solve old crimes. People are studying DNA from old crime scenes and are learning that some people accused of crimes were actually not even present when the crime was committed. This has led to people who have been wrongly accused being released from prison.

RNA

How does DNA help determine things like your eye color and your hair color? Well, DNA's main job is to tell your cells what proteins to make and how to make them. You see, nearly everything that happens in your body is under the control of proteins. By telling your cells what proteins to make and how to make them, DNA is basically telling your cells what to do.

How does DNA instruct the cell when it comes to making proteins? After all, the DNA is in the nucleus. How does it instruct the cell from in there? Well, it does this via a messenger, and that messenger is RNA. RNA copies part of the information that is in DNA and then leaves the nucleus through one of the pores in its membrane. Remember the ribosomes? That's where the RNA goes after it leaves the nucleus. The information carried by the RNA is a recipe for a protein. The ribosome reads that recipe and makes that protein. So the information the cell needs about what proteins to make and how to make them is in the DNA, which is in the nucleus. When DNA wants to instruct the cell, it sends its messenger, RNA, out of the nucleus with a recipe. The ribosome reads that recipe and "cooks up" the protein that the cell needs.

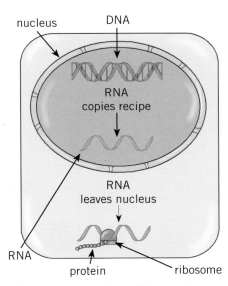

RNA copies recipes from DNA and leaves the nucleus so the ribosomes can make proteins according to the recipes.

Wow. That was a lot of information about cell anatomy. You are becoming quite an expert. Tell someone all that you remember from the last part you read.

Cell Creation

Your body contains billions and billions of cells. Where did all those cells come from? Well, they were copied from the first cell that made you. Every single cell in your body came from one original cell, which was formed at the moment of your creation.

Have you ever tried to copy something? Maybe you do copywork exercises in school. If so, you understand that it's hard not to make a mistake. Well, can you imagine copying something 100 times without making a lot of mistakes? Think of it! As each of us grows from one cell to the many billions of cells that make up a person, the nucleus (with the help of the centrioles) directs the cell to copy itself billions of times! For most

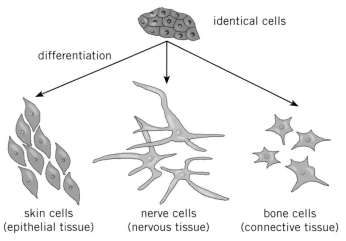

When cells differentiate, they become specific cells that can make specific tissues.

people, all this happens without mistakes.

At first, each cell makes an exact copy of that first created cell. But by the end of the first week of your little life, the cells begin to **differentiate** (dif' uh ren' she ayt). What do you think that means? That means they become different. They differentiate into cells that will go on to become muscle cells, brain cells, and so on. Then, these collections of specific cell types form **tissues**. A tissue is simply a group of cells of the same type.

There are four main kinds of tissue: **nervous tissue**, **muscular tissue**, **connective tissue**, and **epithelial** (ep' uh thee' lee uhl) **tissue**. Muscular tissue makes muscles. Nervous tissue makes up your brain, spinal cord, and nerves, which help you feel the world around you and control the parts of your body you want to control. Connective tissue is tissue that connects to other organs in your body. For example, connective tissue connects your muscles to your bones and keeps your skin where it's supposed to be. The last type of tissue is the kind you see with your eyes. In fact, your epithelial tissue is showing right now, because your skin is made of epithelial tissue! Epithelial tissue also includes the lining of your mouth and nose as well as the lining of many organs. God uses these four tissue types over and over, in different locations and combinations, to form all your many organs.

An organ is a group of tissues that work together to perform a special function. As you develop, your tissues come together to form organs. They begin doing this within three weeks of your life as your body forms inside your mother. In 21 short days, that very first cell has reproduced and has begun to create your brain, heart, and other organs. Groups of organs then begin to form entire organ systems, such as the digestive system, which includes your mouth, stomach, and intestines.

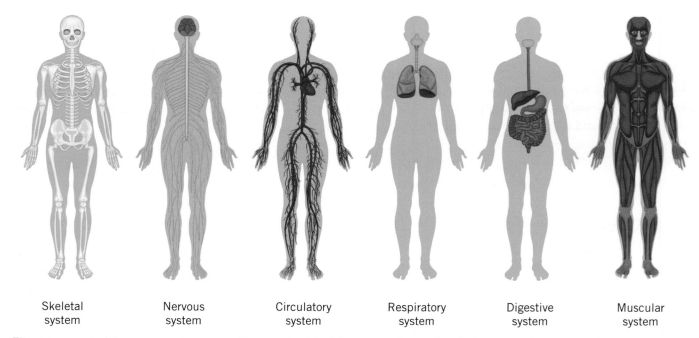

| Skeletal system | Nervous system | Circulatory system | Respiratory system | Digestive system | Muscular system |

These are some of the organ systems you will learn about in this course. Remember that organ systems are made of organs, organs are made of tissues, and tissues are made of cells.

Your body has eleven organ systems, and this book will take you on a journey to learn about most of them. You'll learn about the cells, tissues, and organs that form the organ systems inside your body. Since cells are the building blocks of all systems, you'll get to learn a lot about different kinds of cells as you study anatomy. Cells may be small, but they are very interesting and important.

Even after you have finished growing, every single hour of every single day, your body produces nearly exact copies of millions and millions of cells in order to replace old and worn out cells. For example, did you know that you shed about four pounds of skin cells each year? Count up how many pounds of skin comes from all your family members each year by multiplying 4 times the number of people in your home. No wonder your house gets dusty! As each of these skin cells dies and flakes off, it is replaced by a new cell. Every time you get a cut, new skin is made to replace the damaged or lost cells resulting from your cut. Wow! Your cells are really busy aren't they?

I know you think all cells are simply too small to see with your own eyes, but there are some cells in your body that are actually pretty long. The nerve cells in your legs are thinner than the strands of a spider's web, but they can be more than 3 feet in length! Another kind of cell that you can see is a chicken egg. The entire egg actually starts out as just one single cell.

So now you are a cell expert. Hopefully, you've enjoyed learning all about cells. It's now time to answer some questions about what you've learned. Then, you can do some special activities and projects to have fun learning more about cells!

What Do You Remember?

What tells us that the Egyptians understood a lot about anatomy? How do the laws that God gave to the Hebrews show us that God cares about our health? What was wrong with the way the Greeks decided on their scientific beliefs? What did Galen use to treat gladiators' wounds? What did Hooke call the tiny rectangles he saw in the cork he examined under a microscope? Name the different cell parts about which you've learned.

An egg is a single cell. If a baby chick starts developing in the egg, however, it quickly becomes more than one cell.

Notebooking Activities

You will create a notebook of all you've learned in this study of anatomy. You can purchase a beautiful *Anatomy Notebooking Journal* that goes along with this book, or if you prefer, you can make your own notebook. Simply use a notebook and copy paper to complete the notebooking journal assignments that you will be given in each lesson. Today's notebooking assignment is to write a simple description about the history of anatomy, from the ancient Egyptians to the discovery of the microscope. You can draw or paste a photo about each scientist or discovery. Remember that your cell illustration needs to be put into your notebook as well.

Personal Person Project

Throughout this course you will add pictures of the organs you've studied to a paper model. This will be your Personal Person Project, and it will be kept in your notebook. You will begin by drawing the form of a person on an 8-inch x 10-inch sheet of paper. Then, you'll personalize the form with a picture of your head on top. As you progress through the lessons, you will add the different organ systems about which you are learning to your Personal Person. If you have the *Anatomy Notebooking Journal*, all the templates for completing your entire Personal Person can be found in the appendix. You will only need to add your head to personalize it. You will keep your "progressing" Personal Person on the first page of your *Anatomy Notebooking Journal*. If you do not have the *Anatomy Notebooking Journal*, you can simply draw your version of the organ systems (based on the pictures provided in the textbook) and add them to your Personal Person. You can also print pictures of the organ systems from the Internet. Today, you are going to create the outline of your Personal Person.

You will need:

- A sheet of 8-inch x 10-inch paper (flesh-colored construction paper is recommended)
- A pencil
- Scissors
- A photograph of your face (between 2 and 3 inches tall)
- Tape

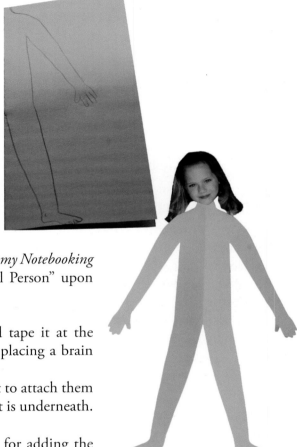

1. Fold your paper in half along the longest side.
2. Reduce the size of your paper by cutting three inches off the top.
3. On your paper, trace half a human body from the shoulders to the feet, beginning at the top folded edge. Use the pictures on the right to guide you in your drawing. Your outline does not have to be perfect.
4. Cut out the half-figure and unfold it to reveal a whole person.
5. Place your person in your notebook. If you have the *Anatomy Notebooking Journal*, there's a page near the front entitled, "Personal Person" upon which you should glue your person.
6. Cut out your face and neck from the picture of yourself.
7. Place the picture on the shoulders of your person and tape it at the top. You want the head to lift up, because you will be placing a brain underneath it later on when you study the brain.
8. As you add organs to your Personal Person, you will want to attach them in such a way that you can lift up the organs and see what is underneath.

At the end of each lesson, you will find instructions for adding the organs you've studied to your Personal Person.

Project
Edible Cell

You learned a lot about cells. It's pretty easy to forget all that you learned. So, you are going to make an edible Jell-O cell to help you remember the different parts of a cell. You will be using various kinds of candy to represent your organelles. Options are given below for some of the organelles. Choose whichever kind of candy you would like to use.

You will need:

- A sharp steak knife and a parent to use it
- A spoon
- A plate
- A glass or ceramic cereal bowl (You'll need one for each cell you intend to make.)
- Cooking spray, like Pam
- A box of yellow colored Jell-O (This will be the cytoplasm.)
- A box of unflavored Knox Gelatin (This will be used to keep your cell together as you add organelles.)
- A jelly bean or a peanut M&M candy (This will represent a mitochondrion, which is the singular of mitochondria.)
- Several Skittles, Everlasting Gobstoppers or M&M candies (These will be the lysosomes.)
- A Starburst Gummiburst or several Smarties (These will make the Golgi body.)
- A Fruit Roll Up (This will be the endoplasmic reticulum.)
- Nerds or cake sprinkles (These will be the ribosomes.)
- Tubular cake sprinkles or Twizzler Pull and Peels (These will be the centrioles.)
- A large gumdrop, jaw breaker, or round chocolate truffle (This will be the nucleus.)

1. Mix the Jell-O according to the package directions, but add one package from the Knox gelatin box and one extra cup of cold water.
2. Spray cooking spray on the inside surface of your bowl.
3. Pour your Jell-O/Knox mixture into your bowl. If there is any left over, do whatever you want with it. Remember, you could make more cells.
4. Let the Jell-O/Knox mixture (the cytoplasm) harden for several hours (overnight is preferred).
5. Once the cytoplasm is hardened, use a sharp steak knife to cut out small segments of the Jell-O, where each organelle will be placed. Be certain not to cut all the way through. Try to make your cut match the size of the organelle you will place in the cytoplasm.
6. Once you have finished placing your organelles into the cytoplasm, carefully turn the bowl over onto a plate. You've just made an edible cell!
7. Take some photographs of your cell for your notebook. Eat and enjoy!

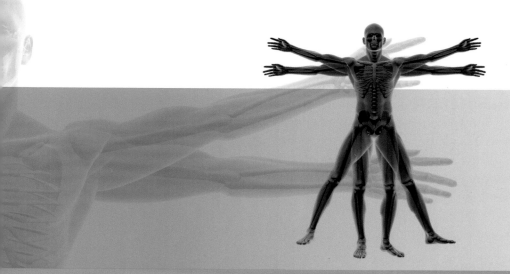

the
SKELETAL SYSTEM

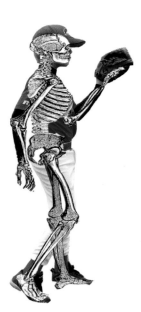

I'm sure you've seen a picture of a skeleton many times in your life. It can seem scary until you consider the fact that a skeleton spends all day with you – even at this very moment, there is a skeleton with you! A skeleton (just like one in a book or museum) never leaves you alone, no matter where you go or what you do. It's inside you all the time. If you removed all your skin and muscles, you would be mostly skeleton. Of course, you wouldn't enjoy life much without your skin and muscles, but it's true that a creepy skeleton is inside you. If that makes you shake a little, let's just imagine life without a skeleton. If you had no skeleton, you would flop around like an octopus. You would have no form or structure; you would look like a blob on the ground. Let's see what you would look like.

Your skeleton supports you so you can stand up, play sports, and do many other things.

Try This!

Shape a human figure from a lump of clay. Give it long, thin legs and arms; put a large head on its shoulders. Go ahead and squish the head flat because that's what your head would look like if you did not have a skull. Now see if the figure can stand up on its own. If you made your figure correctly, it will not be able to stand.

Next, break a few toothpicks into different sizes. Insert the toothpick pieces inside your human figure like bones. You will probably need to rebuild your figure, inserting the toothpicks as you add each body part. Compare your second human figure to your first one. Did the toothpicks help the second figure stand up? They should have. The toothpicks acted like bones in the clay figure's skeleton. They gave it support so it could stand.

A clay figure needs a "skeleton" of toothpicks in order to stay standing.

So aren't you glad you have a skeleton inside you? It's a good thing you do, because your skeleton is an amazing creation of God! The bones in your body do more than just stand you up and keep you looking good. God made the **skeletal** (skel' ih tuhl) **system**, the network of bones inside of you, to do lots of other important things. You could never take out all your God-given bones and replace them with manmade bones, because the manmade "look-a-likes" could never perform all the mind-boggling duties of a single bone. Besides keeping you from flopping around like an octopus on the ground, let's see what else your bones do.

What Do Bones Do?

In addition to supporting you and giving you your shape, bones also protect some very delicate organs in your body – ones that you would want to have protected – like your brain, your lungs, and your heart. Think about it: without your **skull**, every little bump on your head would damage your brain, and you wouldn't be able to think very well anymore. If you fell too hard, well, your brain would simply get mushed. That would be terrible! Your skull, though it may not be pretty, is absolutely one of the most important groups of bones in your body.

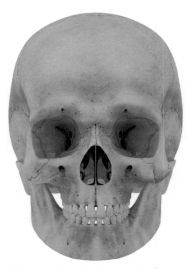

Your skull protects your brain.

Try This!

God was wise to create you with a thick skull to protect your brain, but He didn't stop there. God also put a special layer of fluid between your skull and your brain for added protection. Let's see how this protection works. You will need: two eggs, a plastic container (with a lid that seals tightly) that's only slightly bigger than the eggs, some water, and some masking tape. In this experiment, the eggs are going to represent a brain. Place the first egg in the container. Seal the lid. Jog around the room, shaking the container. What happened? The egg went splat, didn't it? Now wash out the container and try the same thing with the second egg, but this time you will add some protection. Fill the container to the very tip top with water. Place the egg in the container and seal it closed with some masking tape. Now jog around the room in the same way as before. Carefully open the container. What happened to the egg? Hopefully, it didn't break, because the water helped to protect it from damage. Now you know one of the reasons God placed that special fluid between your skull and your brain!

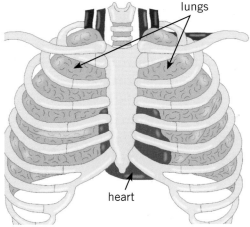

lungs

heart

Your rib cage protects your heart, lungs, and other organs.

You're beginning to realize how God designed some of your bones to protect you from harm. Did you know that if you didn't have a special cage around your heart and lungs, even a small fall could collapse your lungs or make your heart stop working? Do you know what the cage is called? It's called your **rib cage**. We'll go over the bones of your rib cage (and many others) in just a little while.

Got Blood?

So bones give your body shape, and they protect important organs. Do they do anything else? Absolutely! Do you think blood is an important thing to have? Let me assure you, it really, really is. You'll learn all about blood in a later lesson. Have you ever wondered where all the blood in your body comes from? If you bleed, your

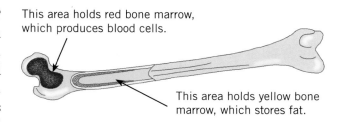

body has to make more blood. How does it do that? Amazingly, your *bones actually make your blood cells.* Can you believe that? Bones are like little blood cell factories. Without your bones, your body would run out of blood pretty quickly, because blood cells don't live very long. So, now do you see why you couldn't take out the bones God made and replace them with manmade bones? God designed the bones to do a lot more than meets the eye!

This area holds red bone marrow, which produces blood cells.

This area holds yellow bone marrow, which stores fat.

Blood cells are made in the red bone marrow found in many bones, and fat is stored in the yellow bone marrow.

Warehouse Wonder

Bones have another function: storage. Like warehouses, they store important things your body needs, such as fats, which scientists call **lipids** (lip' idz). A lot of people try to limit the amount of fat they eat in order to lose weight. However, without an adequate amount of fats, your body would not work properly. That's why your bones store them. Bones also store **minerals** (min' ur uhlz). Yes, just like the earth, you have minerals stored deep inside you. Your body actually uses them to function. For example, one mineral, **calcium** (kal' see uhm), helps to keep your heart beating regularly. It also helps your mind to think and understand the world around you. One time, my grandmother began to forget a lot of things. She forgot my name. She forgot how to make a cup of coffee. We thought she was in real trouble. The doctor found out that she wasn't getting enough calcium. So, they gave her calcium and guess what? All of a sudden, she was back to normal. At the time, none of us knew how important calcium was. We must have calcium to be healthy, so we can be thankful that God created our bones to store it.

Drinking milk is a good way to get the calcium you need.

Even though your bones store calcium, you need to take in some every day. If you don't, your body takes the calcium stored in your bones and uses it to help your heart beat, your brain think, and many other important things. Can you guess what happens to your bones if your body takes too much of their calcium stores? Well, your bones become weak and can be easily broken. Calcium (in the form of a hard material called **calcium salts**) makes your bones strong and sturdy, so make sure you get some calcium every day!

Do you know what foods contain calcium? The main ones are yogurt, cheese, and milk. In fact, three glasses of milk a day will meet most of your body's calcium needs. There are other sources of calcium like spinach, salmon, and ice cream, but milk has more calcium per serving. It's important to get as much calcium as you can today, because about 90% of the calcium stored in your bones is put there by the time you turn 18 years old. People who don't get enough calcium when they are young often develop a disease called **osteoporosis** (ahs' tee oh puh roh' sis). This disease results in weak bones with lots of tiny holes in them. Many people with this disease have bones that break very easily. A fall that would not hurt you very much could actually break one of their bones!

Bone Brawn

Since we're on the subject of bones, let's talk about strong bones for a minute. Having strong bones is very important. As you have probably already guessed, strong bones don't break as easily as weak bones. Besides calcium, what makes your bones strong? Well, one thing is exercise. The more you use your bones, the stronger they become. Bones that are not used become weak and frail. This can happen to astronauts in space.

Without gravity, they just float around, not exercising their muscles or bones much at all. If they didn't do something about that, their legs would become too weak to support their bodies when they returned to earth after a long mission in space. To help avoid this, they strap themselves to machines and run, but they still lose some bone, because running without gravity isn't nearly as hard as running with gravity.

There's something else besides calcium salts and exercise that makes your bones strong: it is a nutrient called **vitamin D**. You can get this from the things you eat and drink, but your body also *makes* this nutrient when you get out in the sunlight. If children do not get enough vitamin D, they can get a condition called **rickets** (rik' its). Rickets is common in countries where food is scarce and children are hungry. It also occurs in the United States when children are indoors a great deal and do not get enough vitamin D from what they eat and drink. Rickets, like osteoporosis, causes bones to be very weak and easily broken. In addition, the bones don't form properly. A child with rickets will often have deformed bones, like leg bones that become severely bowed. When a person doesn't get enough of the right kinds of nutrients, we say that the person is suffering from **malnutrition** (mal' new trish' uhn).

Your bones actually store other minerals in addition to calcium because calcium is not the only mineral you need for a healthy body. One of the other minerals your body stores is phosphorus (fos' fur us). You can get phosphorus from whole-grain cereals, fish, milk, and green vegetables.

Astronaut Jeffery N. Williams aboard the International Space Station is exercising so his bones don't get too weak while he is in space.

Let's Get Moving

The last thing bones do is probably one of the most important to you. Your skeleton is composed of many **joints**, which allow you to move. Of course you need muscles to move the bones in your skeleton, but without your joints, your muscles wouldn't do you much good at all. For right now, however, we are going to concentrate on your bones. We'll talk about your muscles in the next lesson. I know that you probably like to move around and so do I. So, I'm going to keep my skeleton. How about you?

Tell someone what you have learned so far. Include all the things bones do for your body.

Your skeleton has joints in it, which allow you to move your body in many interesting ways.

Bone Anatomy

Have you ever been to a museum and seen skeletons of animals or people? Actually, your bones don't look *exactly* like those skeletons. That's because your bones are alive, and living bones look different from dead bones. How are your bones alive? They're alive because they're made up of living material – material that is constantly working, growing, and repairing itself all the time. Bones are made of cells! Bone cells, to be exact! Let's look at the anatomy of a bone.

On the Outside

The outside layer of a bone is a thin, tough membrane called the **periosteum** (pehr' ee ahs' tee uhm). As you can see from the drawing below, it is filled with nerves and blood vessels. The nerves sense pain. Yes, bones have feelings too! When someone breaks an arm, it's the periosteum that's screaming, "Ouch!" The blood vessels take nutrients to the bone and take out the trash – the waste made by the bone cells. The periosteum also helps in the building of new bone. Did you know your body is constantly building new bone and destroying old bone? Scientists call this **remodeling**. It's like when you play with Legos. You tear down part of a Lego structure and build other parts onto it. Sometimes this makes the structure better and stronger; other times it doesn't. It's the same with bones. How well your bones get remodeled depends on what you eat and how much exercise and sun you get. God has created an amazing system that allows for your bones to heal and become strong, but you have to eat the right foods to give your body the materials it needs to do this. We'll discuss how that happens in just a minute.

Made to Last

The next layer of a bone is the thick, hard layer called **compact bone**. When you see a skeleton or a single bone, this is mostly what you are looking at. The living layer of periosteum has died, dried up, and peeled off the bone. This leaves just the compact bone on the outside. Compact bone is so durable that it can last for thousands of years if preserved correctly. It is smooth and hard, made of many layers of calcium-rich minerals and a tough fiber called **collagen** (kol' uh juhn). These materials are woven together to make bone the second strongest material in your body (you'll learn about the strongest material in Lesson 4). In fact, very few things on earth are as strong as bone. Have you ever held a steel pipe in your hands? If you have, you know that a steel pipe is unbendable. It is so strong that builders use it to create skyscrapers. Of course, it is really heavy, too. Pound-for-pound, however,

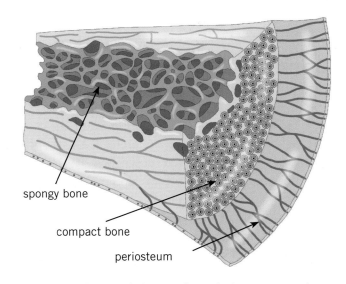

spongy bone

compact bone

periosteum

This bone has three main layers: the periosteum, compact bone, and spongy bone. Bone marrow fills the spaces between the spongy bone.

your bones are actually stronger than steel! But even more amazing is the fact that, though bones are stronger than steel (on a per-pound basis), they are still fairly flexible. That's partly because God placed something called "spongy bone" on the inside.

Bouncy Bone

Under the compact bone layer is a network of pores and tunnels interconnected in a pattern that makes the bone strong yet resilient (rih zil' yent – able to bounce back after being compressed). This network of bone tunnels is called **spongy bone**. Why did God design our bones to be resilient? Well, have you ever had anything fall on you? One time, when I was young, a car door slammed on my hand. Ouch! That really hurt. But it didn't break my hand! Do you know why? Because the bones in my hand were resilient enough to compress and then bounce back without breaking. God designed our bones to withstand a certain amount of pressure, because He knew that sometimes our bones would have some pressure put on them.

In the Marrow

Do you have any idea what is on the very inside of many of your bones? If you break open a fresh chicken bone, you'll find out. There's a hollowed-out cavity with some fluid inside – just like in one of your bones. You see, on the inside of many bones is a cavity filled with a thick fluid called **bone marrow**.

There are two kinds of bone marrow: red and yellow. **Red bone marrow** is the kind in which blood cells are made. When you cut yourself and bleed, your body needs to replace that blood. Well, your red bone marrow helps your body make more blood by providing it with new blood cells. This is important even if you don't cut yourself, because some of your blood cells (the red ones) aren't very hardy. They die of old age after only four months, so the red bone marrow must constantly produce new blood cells to replace the ones that are dying.

The **yellow bone marrow** stores up lipids. Do you remember what lipids are? Fats. These fats are used throughout your body to perform many important functions. In fact, the cell membrane you studied in the first lesson contains a lot of fat. Without fat in your diet, there would be no cell membranes. Without cell membranes…there would be no cells. No cells would mean no you. Thank God He made fat!

Now you know about bone anatomy.
Tell someone everything you can remember before moving on to the next section.

Bone's a Growing!

Now put on your thinking cap because we are going to get a little technical here. There are some really important cells living in your bones. They all begin with "**osteo**" (ahs' tee oh), because "osteon" is a Greek word that means "bone." The cells that make new bone are called **osteoblasts** (ahs' tee oh blasts'). Osteoblasts lay down new layers of bone, giving you more bone. How do they do it? Actually, they use a few ingredients that you provide! They need you to get plenty of minerals (like calcium) and exercise. They also need protein to make collagen, and they need vitamin D. Your body can make vitamin D when your skin is exposed to sunlight, or your body can get it from some foods. With all the right ingredients, those osteoblasts are having a blast making your bones bigger and stronger!

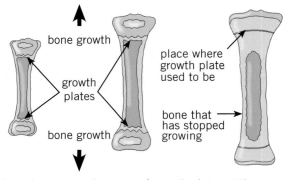

Long bones grow because of growth plates at the ends of the bone.

What if your bones never grew any longer from the time you were a toddler? You would be super short. Bones grow longer in special places called growth plates. On long bones (like arms and legs), growth plates are near the ends of the bone. So, bones don't grow from the middle, they grow from the ends.

This is the way it happens: when you go through a growth spurt (children grow in spurts as well as gradually over time), a substance called **cartilage** (kar' tuh lij) begins to stack up on the growth plate. Feel the tip of your nose. It has no bone; it is made out of cartilage. Cartilage is firm but resilient and absorbs shock very well. God also put cartilage on the ends of bones that come into contact with each other. Working like a shock absorber, that **joint cartilage** keeps your bones from getting hurt when they rub against one another. Try this experiment to help you understand.

Try This!

Go to your kitchen or bathroom counter top and slap it with your hand - but be careful not to slap it too hard! Now place a pillow or a cushion on the counter top and slap it again. Did the cushion help you from feeling the shock of your hand slapping against the hard counter? That's part of what cartilage does between your bones.

So, you have joint cartilage between your bones at your joints, but you also have growth cartilage that forms your bones' growth plates. When your bones need to grow, cells in the cartilage of the growth plate make the growth plate get thicker. Soon, when it has a big enough stack of cartilage, the cells that were producing the cartilage die, and osteoblasts come to take their place so they can turn the dying cartilage into bone. As a result, you get taller! Your arms and legs get longer! Before you know it, you're looking *down* at mom instead of *up* at her!

Believe it or not, you can actually get an injury to your growth plates that can stop your bones from growing. When my pastor was a teenager, he accidentally stepped into a hole while running in the woods. He broke the growth plate in his tibia, one of the bones in his leg. That part of his leg quit growing, while the rest of his body (including his other leg) continued to grow. His other leg grew two inches longer than the leg he injured! This isn't as likely to happen today, because modern medicine can treat such bone injuries and help the bone grow normally.

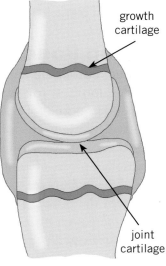

growth cartilage

joint cartilage

When bones rub against each other like they do in this knee joint, the joint cartilage at the end of each bone acts as a firm cushion.

Deep and Wide

In addition to growing longer, your bones also grow wider. Have you ever noticed that your wrist is a lot smaller around than your dad's wrist? Well, that's because bones get wider as they get longer. Here's the way it happens: the osteoblasts lay down new layers of spongy bone right under the periosteum. This spongy bone is then "filled in" to make compact bone. At the same time, different bone cells, called **osteoclasts** (ahs' tee oh klasts') eat away at the spongy layer on the inside of the bone. This widens the cavity inside the bone as the bone gets wider. This kind of bone growth happens even more when you exercise, as long as your body weight isn't too low and you eat plenty of healthy foods.

Have you ever wondered when you'll stop growing? If you're a boy, you'll keep growing in height until you're about 18 years old. Girls stop growing in height around the age of 16. But here's something interesting to think about: at nine years old, a boy is 75% of the height he will be as an adult. On the other hand, a girl is 75% of her adult height at seven years old. So girls grow a bit faster and stop growing a bit earlier than boys.

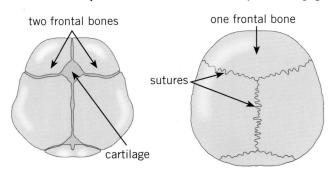

two frontal bones

one frontal bone

sutures

cartilage

There are five visible bones on the top of the infant skull shown on the left. They have fused to four bones in the adult skull on the right.

Bones do more than just grow longer and wider as you become an adult. Believe it or not, a baby is born with about 300 bones, but an adult has only 206. Wait a minute! What happened to all those baby bones? I'm going to let you take a guess by looking at drawings of a baby skull and an adult skull. Do you see what happened to the bones from infancy to adulthood? The bones **fused** together. In other words, two (or more) smaller bones became one bigger bone. This happens when the cartilage between the bones turns into bone, joining them together.

Even once your body has finished growing, your bones are being constantly remodeled. Remember the Legos we talked about? They're always being reconstructed and rearranged. You see, under the stress of everyday activities (walking, running, biking, etc), your body reshapes your bones so they can give you the best support for the kinds of activities you do most often. Even as you sit reading this book, calcium and phosphorus are being removed from some areas of your bones and deposited into others. This is an amazing design God chose for our bones! Let's see why.

Broken Bones

Have you ever thought about how amazing it is that God's creations repair themselves, while man's creations typically don't? If you break a plastic leg on a toy, it pretty much stays broken. But that's not the case with the wonderful body God created! Amazing things happen when you injure yourself. In fact, your body is specifically designed to fix itself. Even broken bones fix themselves! You may be wondering why people have to wear casts if bones just repair themselves. Well, the cast holds the bones in place and sometimes pushes the two broken pieces of bone closer together so that healing happens faster. You've probably noticed that people who are badly hurt try to keep the injured area of their body from being jostled. A cast serves the same purpose – it keeps the broken parts of the bones from moving. The support it provides promotes healing and helps relieve the pain a broken bone can cause. Even ancient people understood that keeping broken bones from moving decreased pain and also sped up the healing process. Before modern casting materials were invented, broken bones were kept still with wood splints wrapped in cloth, or tree bark, or linen stiffened with a variety of items including egg whites, flour, starch, or wax. Eventually, people began using plaster of Paris. For a long time plaster was the standard casting material, but now most doctors use fiberglass, because it is lighter, more durable, and faster to apply.

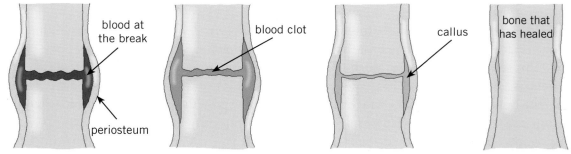

A broken bone heals by forming a blood clot that allows osteoblasts to come in and make a callus, which is then remodeled into normal bone. The new bone is nearly the same as the old one, but might be a bit thicker at the point of the break.

So, how does a bone heal itself? First think about what happens when you cut your skin – a scab forms, and eventually, new skin grows back. This process starts with a clot of blood forming at the site of the injury. The same thing happens when a bone is injured. When you break a bone, it bleeds. The blood settles between the bones and makes the periosteum bulge out. Then a clot forms. Osteoblasts come into the clot and start making new bone. The bone fills in the space between the bones but also the bulge under the periosteum. This is called a **callus** (kal' us). Over time, the osteoclasts absorb the callus, and the osteoblasts produce bone that is very similar to the original. When you put a cast on a broken bone, it keeps the bone ends close and the callus unbroken so your body can remodel that bone more quickly and get it back in action.

Take some time to tell someone what you learned about bone growth, remodeling, and healing.

Bone Basics

How do you suppose the bones in your skull are different from the bones in your fingers? Although all bones in your body are made of the same material, they are obviously different shapes and sizes. Can you guess which is the longest bone in your body? Look at the skeleton on the right to find your answer. Did you figure out that it's your thigh bone? That bone is called your **femur**. Don't worry, in a little while we'll do an activity to help you learn the names for many of the bones in your body. Do you know which is the smallest bone in your body? There are three teeny tiny bones in each ear. All three of them could fit on a dime with lots of room to spare. They are called the **malleus** (mal' ee us), **incus** (ing' kus), and **stapes** (stay' peez). The smallest one is the stapes, measuring 2.5 mm long. Use your ruler to see how long that is.

Shapin' Up

As you can see, your bones have many different shapes, and scientists have given names to them. For example, **long bones** are – well, long. They are longer than they are wide. We've already discussed one long bone, the femur (your thigh bone). Other long bones are found in your legs, arms, fingers, and toes.

      Short bones are…you guessed it…short. They're usually around the same height and width. Short bones are found in the ankle and wrist. **Flat bones** are another kind of bone. The bones in your skull are considered flat bones. They are not round and thick like the bones in your legs. Your sternum and ribs are also flat bones.

      Sesamoid bones are small and usually rounded, like small stones. There are some tiny ones in your hands, but the easiest one to find is the **patella**, or knee cap. Stand up and bend over to feel your knee. Feel around until you find a bone that moves back and forth. That's your patella! The rest of the bones, like your vertebrae, are called **irregular**, because they don't fit any of the other categories.

Connect the Bones

Have you heard the bones song? When I was a kid, we used to sing this song a lot. Here's a part of it:

    Ezekiel cried, "Dem dry bones!"
    "Oh, hear the word of the Lord."
    The toe bone connected to the heel bone,
    The heel bone connected to the foot bone,
    The foot bone connected to the leg bone,
    The leg bone connected to the knee bone,
    The knee bone connected to the thigh bone,
    The thigh bone connected to the back bone,
    The back bone connected to the neck bone,
    The neck bone connected to the head bone,
    "Oh, hear the word of the Lord."

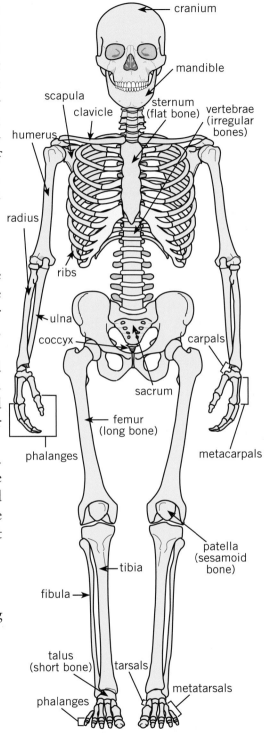

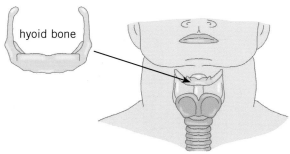

The hyoid bone is connected to cartilage of the larynx (lar' ingks), but not to any other bone.

This fun song teaches an important truth. You see, every bone in your body is connected to another bone. Well, that's not entirely true. There is one lone bone in your neck, called the **hyoid** (hi' oyd) bone, that isn't connected to any other bone. It is connected only to cartilage.

Ligaments

Bones are connected to one another by special tissues called ligaments. These tissues are made of tough collagen fibers that hold the bones in place at a joint. They limit the joint's motion so that the bones don't get too far away from each other. Because they can stretch a bit, however, they allow the bones to move farther apart from one another for a brief time before snapping back into place.

Because bones are connected to one another, they can affect each other. Did you know you can get a concussion (an injury to the brain) by falling on your **coccyx** (kok' siks), or tailbone, really hard? It's true. My son was wearing tennis shoes with wheels and fell on his tailbone so hard that it shook his brain and gave him a minor concussion! This may seem hard to believe, but if you bend over and run your fingers along your spine, you will feel the ridges all connected to one another – from your tailbone all the way up to your skull. Now you can see how my son's tailbone injury affected his brain, even though it was halfway up his body! This is also why shaking an infant is so dangerous.

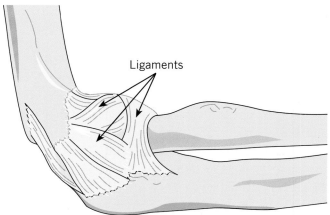

These are the ligaments that hold the bones of the elbow joint in place.

A Head of the Game

The part of your skull that protects your brain is called the **cranium** (kray' nee uhm). Although it feels like it's just one rounded flat bone, it is actually made up of eight bones (once all your infant bones fuse). These are important bones since they cover and protect your brain. Nevertheless, these bones fit tightly together. The place where any bone meets another bone is called a joint. So, even though your cranium does not move, it has joints. Joints that do not move are called **sutures** (soo' churz). We will learn about several other kinds of joints later on.

In babies, the sutures allow for brain growth. You probably remember that the sutures are not fused or sealed together in a baby. This is because the brain isn't its full size yet. As the brain grows, the cranium grows by being gently pushed outward by the brain. This way, the brain is always protected in its coat of armor, but the armor can grow with the brain! That's a marvelous plan designed by a marvelous God! Eventually brain growth stops, and the places where the bones of the cranium meet become fused at the sutures.

occipital bone (ahk sih' puh tuhl)

parietal bone (puh rye' uh tuhl)

temporal bone (tem' puh ruhl)

frontal bone

maxilla (max sil' uh)

mandible (man' duh bul)

zygomatic bone (zye guh mat' ik)

There are nearly 30 bones in an adult's skull.

Let's Face It

Moving down from the cranium, you'll find your lovely **facial** (fay' shul) **bones**. Of the 14 bones in your face, all except one are attached to the other bones with a suture joint. In all those bones, only one bone moves. Can you guess which bone that is? Well, open your mouth really wide. Now do you know? Look at the drawing and tell me the scientific name for that bone. It's the **mandible**!

The malleus, incus, and stapes are so small they can fit on a dime, with room to spare!

The three smallest bones in your body are located in the skull. Do you remember where? The ear! Do you remember their names? They are the malleus (sometimes called the hammer), incus (also called the anvil), and stapes (or stirrup). The alternate names I put in parentheses come from their shapes. Can you tell which is which in the picture on the right?

If you were to look into some of the facial bones, you would find hollowed-out cavities called **sinuses** (sye' nuh sez). These "holes in your head" make your head lighter, and they are lined with special cells that produce **mucus** (myoo' kus). If you get a bad cold and your nose runs, it is probably the result of mucus draining from your sinuses, which empty into your nose.

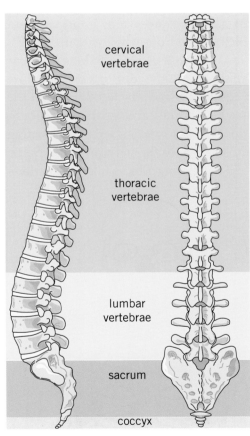

The side view (left) and rear view (right) of the spinal column show how the vertebrae stack together.

Shivers Down Your Spine

If you studied zoology, you learned that you are a **vertebrate** (vur' tuh brayt), meaning you have **vertebrae** (vur' tuh bray). What does all this mean? It just means you have a spine, which we call your **spinal column** or **vertebral column**. How many times in a row can you say vertebral column? I can only say it twice before I mess up.

Your spine is made up of all those hard bumps going down your back. Those bumps are donut-like bones called vertebrae. Can you count how many bumps you have? That's hard to do. Try it. Start right behind your neck and count each bump. Your neck has seven vertebrae. They are called your **cervical** (sur' vih kul) **vertebrae**, because "cervical" refers to the neck. There are twelve **thoracic** (thu ras' ik) **vertebrae** ("thoracic" refers to the chest) and five **lumbar** (lum' bar) **vertebrae** ("lumbar" refers to the lower back). That's a lot of bumps in your back! The spinal column is designed like this because inside the donut hole of each of those vertebrae is your **spinal cord** – coming from your brain and going all the way down your back. The spinal cord is an important part of the nervous system, which allows you to feel, walk, move, and talk. It's an important thing to protect.

Because of the way God designed the vertebrae to stack on top of one another, the spinal column can move in many, many different directions and still protect your spinal cord. Strong ligaments keep the vertebrae from falling off one another as the spine moves about. It's really an amazing design!

These 24 vertebrae also have cartilage discs between them to allow for shock absorption. The bottom two bones of the vertebral column, the sacrum and the coccyx, are made of flattened vertebrae that are joined together.

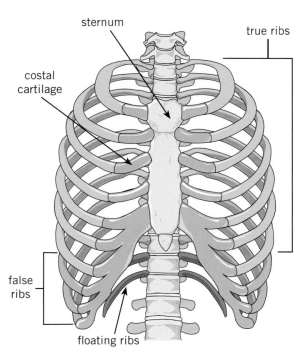

sternum

true ribs

costal cartilage

false ribs

floating ribs

Most people have 12 ribs on each side of the rib cage, and most of the ribs are attached to the sternum with cartilage.

Baby Back Ribs

Most people have about as many ribs as they do thoracic vertebrae. Can you guess why your ribs are so important? I gave it away earlier. Your ribs guard your heart and your lungs (as well as a few other organs). God had to give them a unique "cage-like" design to do this, because your lungs don't stay the same size throughout the day. Put your hand on your rib cage and take a deep breath in and a deep breath out. Did you notice how your rib cage got bigger and smaller with your breathing? That's because in the front, your ribs are attached to your sternum with resilient tissue called **costal** (kos' tul) **cartilage**. This allows your rib cage to expand when you breathe. In the back, your ribs are attached to your thoracic vertebrae.

Look at the diagram of the ribs. Seven of your ribs are **true ribs**, five are **false ribs**, and the bottom two false ribs are called **floating ribs**. Although the false and floating ribs are not considered true ribs, they are just as important and valuable as the true ribs. Let's find out which ribs are which. Look at the top seven ribs, and see if there is anything different about them compared to the ribs below. The first seven ribs are true ribs because their costal cartilage attaches directly to the sternum. The next three have costal cartilage that attaches to the sixth and seventh rib. The last two ribs are called floating ribs because they are not attached to the sternum at all. Some people have an extra set of floating ribs, while others don't have any.

A Peck of Peppers

Have you ever heard people refer to the muscles on their chest as "pecs?" That's short for pectoral (pek' tur uhl) muscles, which are connected to the **pectoral girdle**. Your pectoral girdle connects your arms to your sternum (chest). The word "girdle" means "belt." So, the pectoral girdle is like a belt of bones around your chest and back. It includes the **scapulae** (skap' yuh lay), which are your two shoulder blade bones, and your two **clavicles** (klav' ih kuls), which we often call collar bones. Can you feel your clavicles? They are the bones right below your shoulders that lead inward to your throat. Now reach behind your back and feel your shoulder blades. Those are your scapulae.

Stop for a moment and name all the bones you have learned about so far. Name any others we haven't covered yet that you think you can remember.

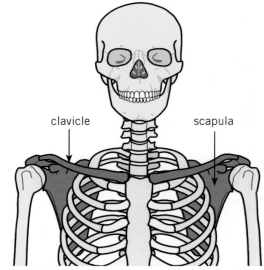

clavicle

scapula

The clavicle and scapula form your pectoral girdle, which attaches your arms to your spine.

Armed and Dangerous

Now, let's study some long bones. Do you remember what classifies a bone as a long bone? It has to be longer than it is wide. The bones in your arms and fingers definitely fall into that category. Let's see if we can locate these bones. Beginning at your scapula, you'll find your first arm bone – the **humerus** (hyoo' mur us). Let's keep going down your arm. Although the upper part of your arm has only one bone, the lower part has two, the **radius** (ray' dee us) and the **ulna** (uhl' nuh). Can you find these two bones? It's easier if you search just above your wrist: your radius can be felt above your thumb, and your ulna can be felt above your pinky finger. Say the names of the bones as you feel them – it will help you remember them.

There are eight short bones in your wrist. They are called **carpals** (kar' pulz). You won't be able to feel them, but you can feel the **metacarpals** (met' uh kar' pulz). The metacarpals are actually hand bones, even though they look like fingers in the drawing. Finally, there are the **digits**. Digits? Well, scientists call fingers "digits." The bones in your digits are called **phalanges** (fuh lan' jeez). Feel your fingers and count the phalanges. You should be able to find three in each digit – except for your thumb, which has only two.

phalanges
metacarpals
carpals

radius

ulna

humerus

scapula

clavicle

Try This!

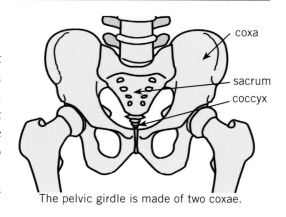

Your arms when spread out are nearly as long as you are tall.

Did you know that you are about as tall as you are wide? That's true when your arms are stretched out at either side. You see, your arms together are about the same length as your entire body is tall. Get a tape measure. Now spread your arms out on either side and measure them from the tip of your middle finger on one hand to the tip of your middle finger on the other hand. Next, measure your height against a wall. The measurements should be nearly the same!

Girdles Around

Moving on down the body, we find that there is another girdle just below your waist! It's your **pelvic girdle**. The pelvic girdle is much like the pectoral girdle, but most of the bones are larger and stronger. God designed them this way because they have to bear a lot of weight and support you as you run, jump, and rock climb – or whatever else you enjoy doing. And of course, those all-important legs attach to the pelvic girdle.

The pelvic girdle connects to your vertebral column and

coxa

sacrum

coccyx

The pelvic girdle is made of two coxae.

contains your hips, sacrum, and tailbone. Do you remember the name for your tailbone? It's your coccyx. Now, put your hands on your hips. Your hips are called your **coxae** (or **coxa** if you're referring to only one of them). When you put your hands on your hips, you put them on the upper part of the coxa on each side.

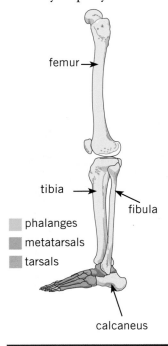

femur

tibia

fibula

phalanges
metatarsals
tarsals

calcaneus

The Last Leg

Like the upper part of the arm, the upper part of the leg has only one bone. It's called the **femur**. Do you remember what is special about the femur? It's the longest bone in your body. Now look at the drawing of the lower part of the leg. It's a bit similar to the lower arm, isn't it? There are two bones there, the **tibia** (tib' ee uh) and the **fibula** (fib' yuh luh).

Learning the bones in your feet is pretty easy. They are a lot like your hand bones. Do you remember that your hands are made of carpals and metacarpals? Well, your foot is made of **tarsals** (tar' sulz) and **metatarsals** (met' uh tar' sulz). You have seven tarsals in your foot, including your **calcaneus** (kal kay' nee us), which is your heel bone. You don't need to learn the other tarsals, but look at the picture to see where they are located. Just like your metacarpals, your metatarsals look like your toes, but they are a part of your foot. Do you remember that your fingers are made of bones called phalanges? Guess what your toe bones are called? They're called phalanges too, just like your finger bones.

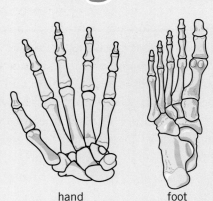

hand   foot

Using the drawings on the left, count how many bones are in the hand and foot. Now multiply each number by two (since you have two hands and feet). How many bones are there? If you counted correctly, you would have found 27 in the hand and 26 in the foot. All together, that adds up to 106. Do you remember how many bones are in your entire body? There are only 206 once you are an adult. That means more than half of your adult bones are in your feet and hands! I guess God considers these important parts of your body! Think about how much you use your hands each day. You need them to create things, take care of yourself, and do your schoolwork, among other things. What about your feet? They are also important. How would you get from here to there, play sports, or dance without them?

**Before moving on, review the bones of the arms, hands, legs, feet, and girdles.
Tell someone else what you have learned.**

Just to get an idea of how important the bones of your hands and feet are, let's make your fingers a little less mobile and see how it feels to do daily activities without them working properly. Use masking tape to fasten your index finger to your middle finger on both hands. Now see how long it takes you to do some of your daily chores. I expect there are some things you can't even accomplish like this. When God designed your fingers, He knew exactly the best way to configure them, didn't He?

Try This! It's hard to remember all the bones we just covered, so we are going to do an exercise that will help you remember them. While looking at the drawing on page 45, touch each part of your body while saying the name of the bone. For example, you will touch the top of your head and say, "cranium." Okay, let's go! Name those bones! Saying the names aloud will help you remember the names of the bones.

Joint Venture

Your bones are wonderfully made, but without joints, you would be a statue – unable to move. Joints are the places where bones meet. Some joints are designed to allow a lot of movement; others allow just a little movement, and still others join the bones together but don't allow any movement at all. Do you remember the joints in your cranium? These suture joints don't move at all. God designed a special fluid to collect in the joints that move. It's called **synovial** (sih noh' vee uhl) **fluid**, and it allows the joints to move more easily.

Without joints, you would be as stiff as this mannequin.

Try This! Rub your hands together really hard for one full minute. Do you feel how warm they have become? The rubbing action creates what we call friction, and friction generates a lot of heat. Now apply lotion to your hands and rub them together like you did before.

It's much easier to rub your hands against one another with lotion on them, isn't it? This gives you an idea of how the synovial fluid helps your joints. God knows all about friction, and He doesn't want that to be a problem for our joints. Without synovial fluid, the cartilage on the end of one bone would rub against the cartilage on the end of the other bone. The friction caused by this rubbing would harm the bones, so God made joints with their own friction-reducing fluid. Elderly people sometimes get a disease called **arthritis** (ar thry' tis). In some forms of this disease, the synovial fluid in their joints is thin and weak, making it painful to move them. Eventually, the bones are remodeled, becoming larger to deal with the extra wear and tear. This further increases the pain associated with moving their joints. Synovial fluid is quite important, isn't it?

Kinds of Joints

There are different kinds of joints in your body, and they offer different ranges of motion. The major ones are:

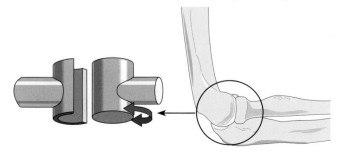

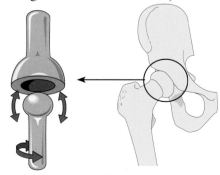

Hinge Joints
Elbows and knees are hinge joints. They offer a limited range of motion but are very stable.

Ball-and-Socket Joints
Hips and shoulders have ball-and-socket joints. They offer a wide range of motion but are less stable than hinge joints.

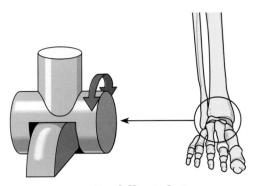

Saddle Joints
The ankle has a saddle joint. The range of motion is less than a ball-and-socket but more than a hinge.

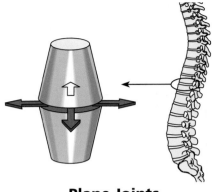

Plane Joints
Your vertebrae are connected with plane joints. They allow you to bend and twist your back.

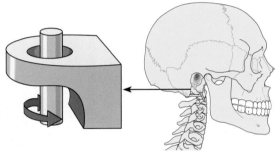

Pivot Joints
When you shake your head "no," you are using the pivot joint that connects your skull to your vertebral column.

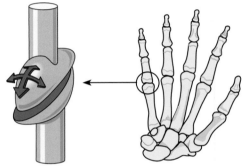

Ellipsoidal (ih lip' soyd uhl) Joints
The joints that connect your phalanges to your metacarpals are ellipsoidal joints. They are like flattened ball-and-socket joints.

What Do You Remember?

Name the different things that bones do for your body. What is the periosteum? What mineral makes compact bone strong and hard? What is the bone tissue that forms tunnels and pores called? What are the two kinds of bone marrow? What are osteoblasts? Where are the smallest bones in your body found? Which is the longest bone in your body? What do ligaments do? What is the rounded part of your skull called? Can you name at least five other bones in your body by their scientific names? Can you name at least one kind of joint?

Notebooking Activities

For today's notebook assignment, find a picture of a skeleton that is not labeled and label all the bones you have learned about. Be sure to write down all the interesting facts you learned on another sheet of paper.

Personal Person Project

It's time to add some bones to your Personal Person. You should add an entire skeleton from the neck down. If you do not have the *Anatomy Notebooking Journal*, you can draw the skeleton on a white sheet of paper and glue it onto your person. If you have the *Anatomy Notebooking Journal*, there is a skeleton in the appendix for you to cut out and use.

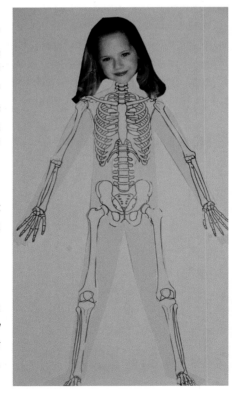

Experiment
Analyzing a Chicken Bone

To better understand what bones are made of, we want you to experiment on a *chicken* bone. If we are studying the *human* body, why are we going to do an experiment with a chicken bone? Well, chicken bones are similar to our own. They are not exactly like human bones, but they are similar enough that we can learn a few things about our own bones by studying those of a chicken. Also, chicken bones are *a lot easier to find and experiment on!*

You will need:
- A cooked chicken wing
- A pair of rubber or plastic gloves
- White vinegar
- Two plastic containers with lids (just big enough for a chicken wing and some liquid)
- Plastic wrap
- A parent with a knife

1. Put on the gloves.

2. Using your fingers, remove all the meat from the bone. If you can't get it all off with your fingers, ask your parent to use the knife to help you. As you are doing this, see if you can see anything else besides meat. The meat is muscle, but the bones are connected with ligaments, and there is cartilage at the end of the bone. Can you see any cartilage or ligaments? Don't worry if you can't. That's not the main point of the experiment.

3. Look at the bones. Notice that there is one thick, long bone that connects to two thinner long bones. Does that sound familiar? It should – that's what the bones in your arm are like!

4. Cut or break the largest bone (remember, it's called the humerus) in half and look on the inside of the bone. Do you see something red there? What is that red stuff? It is the chicken's red bone marrow. That's where the chicken's blood cells come from.

5. Do you remember what material makes the bone strong? It's calcium. We are going to do an experiment that actually removes calcium from the bone to see what happens to it. Fill one of the containers about half full of vinegar.

6. Fill the other container about half full of water.

7. Pull the two thinner bones of the wing away from each other.

8. Put one of the two bones in the container that has water.

9. Put the other bone in the container that has vinegar.

10. Put plastic wrap over each container.

11. Put the containers aside for a few days.

12. Vinegar is an acid that will remove the calcium salts from the bone. Using a Scientific Speculation Sheet, make a hypothesis about any differences you think you will find between the two bones.

13. After three days, check your chicken bones. What happened? Record your results on your Scientific Speculation Sheet.

As you learned while you were doing the experiment, vinegar is an acid that destroys the calcium salts that make bones hard. After a few days in vinegar, then, the one bone lost a lot of its calcium salts. What did it feel like? It should have felt soft and rubbery. You see, bones are a mixture of a great many things, but as you learned in the lesson, collagen and calcium are two of the important ones. Collagen is flexible, and calcium salts are hard. Together, they make bones very strong, but somewhat flexible. When you get rid of the calcium, the bone is no longer hard, but it is still very flexible. The bone in the water might have been a bit cleaner than when you put it in there, but that's about it. It should have felt the same as when you handled it three days ago.

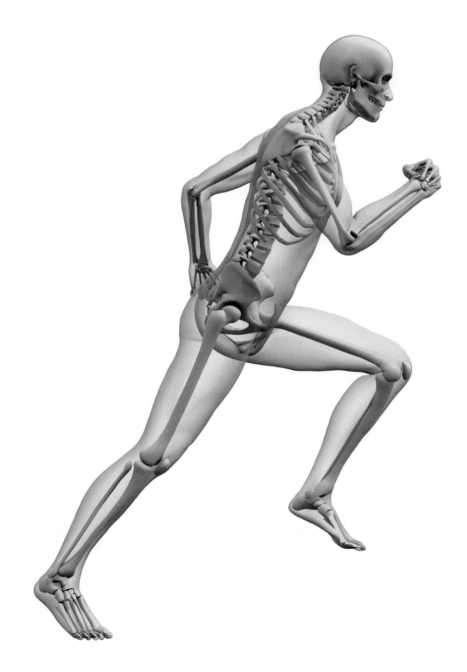

Your bones are strong and flexible because they are made from a
mixture of strong calcium salts and flexible collagen.

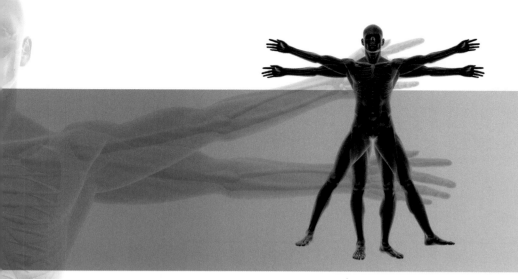

the
MUSCULAR SYSTEM

Make a muscle! You should be very proud of that muscle, because your muscles are a miraculous, marvelous, meaningful invention of a majestic God. You might look like a jellyfish without your bones, but without your muscles, you would be a sack of bones flopped on the ground. In fact, you couldn't move yourself one inch without your skeletal muscles – the muscles attached to your bones. You also couldn't breathe without the skeletal muscle under your rib cage – the one that allows your lungs to expand and take in air. Without the all-important **cardiac** (kar' dee ack') muscle – your heart muscle – you simply wouldn't be able to make it in this world. After all, you gotta have a heart! So, although bones are very important, you really couldn't live without your muscles.

When you bend your arm to "make a muscle," you are flexing your forearm.

Now that we have established how important muscles are, let's explore all the things that make them so marvelous, miraculous, and meaningful to your life. Do you remember how many bones an adult has? About 206. Well, believe it or not, there are a lot more muscles in your body than bones. You have, at this very minute inside your body, more than 640 muscles. That's a big number. In fact, if all the muscles in your body were used at one time with all their power pulling in one direction, they could easily push over a semi-truck. But our muscles don't work like that. They all work in small groups – sometimes even in pairs. Your muscles are also quite heavy. All together they make up about 40% of your entire weight. So, if you weigh 100 pounds, your muscles alone weigh about 40 pounds.

Try This!

Let's find out how much your muscles weigh. You will need a calculator or the ability to multiply decimals. Either will do. Now, put on very light clothes and go step on a scale to see how much you weigh. The lighter your clothes are the better, because you want to measure your weight, not the weight of you and your clothes. If you don't have a scale, you'll have to save this activity for when you find a scale you can use (the doctor's office, health club, or house supply store will have one). Now, multiply your weight by 0.40. What is your answer? That's about how much your muscles weigh.

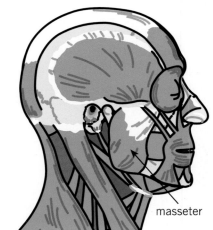

Your muscles make up about 40% of your body weight.

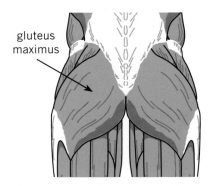

gluteus maximus

Two gluteus maximus muscles form most of your rear.

Just like bones, muscles come in all shapes and sizes. Your smallest muscle is in your middle ear. It holds the teeny tiny ear bones stable. The largest muscle is your **gluteus** (gloo' tee us) **maximus** (mak' suh mus). It connects to the back of the pelvic girdle and the femur. Your gluteus maximus helps you walk, run, jump, climb, and do pretty much every activity you enjoy.

The strongest muscle in your body is the hardest to control and the most dangerous. Read these scriptures and tell me if you can guess where that muscle is located:

- *The one who guards his mouth preserves his life; the one who opens wide his lips comes to ruin.* (Proverbs 13:3)
- *A fool's mouth is his ruin, and his lips are the snare of his soul.* (Proverbs 18:7)
- *With his mouth the godless man destroys his neighbor, but through knowledge the righteous will be delivered.* (Proverbs 11:9)
- *Even a fool, when he keeps silent, is considered wise; when he closes his lips, he is considered prudent.* (Proverbs 17:28)
- *From the same mouth come both blessing and cursing. My brethren, these things ought not to be this way.* (James 3:10)

I'm sure you figured it out right away – the muscle that closes your mouth is the strongest muscle in your body. That muscle is called the **masseter** (ma' sih tur). The masseter muscle actually works to keep your mouth closed. It's a muscle we should exercise more often. In fact, it's dangerous not to use this muscle, because if we aren't careful with our words, we might say things that bring harm to others and ourselves. Maybe the reason God made the masseter to be the strongest muscle is so that we would use it often.

Another reason the masseter muscle must be so strong is because the food we eat is not always soft and tender like baby food. The masseter allows you to bite down on your food and chew it up. It pulls the mandible (remember – you learned about that bone in the previous lesson) upward with a powerful force that can break even bone. In fact, the masseter is so

masseter

The masseter muscle can produce more force than any other muscle in the body.

powerful that your teeth could actually break if you were to bite down on something really hard.

Have you ever heard of a disease called **tetanus** (tet' uh nus)? It's actually an illness that affects the masseter muscle (among others). If you get tetanus, the masseter tightens and won't allow you to open your mouth. Some people call it "lockjaw," because the jaw is locked closed. Have you had a tetanus shot in the last ten years? A tetanus shot protects you from this condition.

Although the masseter is very powerful, it's not the most active muscle in your body. Do you have any idea which muscles are the most active? You might have guessed your heart muscle, and your heart does pump every minute of every day of your life, but you might be surprised to learn that your eyes are even more active than your heart. They move almost all the time! Even when they are closed, your eyes are busy, busy, busy. Even when you're sleeping! Your eyes move so much while you sleep that part of your sleep cycle is called **rapid eye movement sleep**, or **REM sleep**. If you wake up and remember a dream, you probably had that dream while you were in REM sleep.

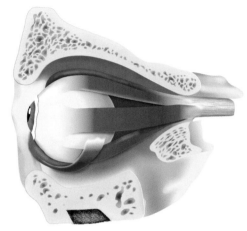

The muscles that control your eye are the most active muscles in your body.

So what kinds of muscles are in your body? Well, you have three different kinds of muscles in your body: **skeletal muscles, smooth muscles** and **cardiac muscle**. Let's have some fun exploring each of these kinds of muscles.

Skeletal Muscles

Biceps brachii contracted: It is short and fat and has flexed the elbow.

Biceps brachii relaxed: It is long and thin and ready to flex the elbow.

Your biceps (bye' seps) brachii (bray' kee eye) can flex your forearm when it contracts.

Skeletal muscles are the ones you typically see when you look at a drawing of the muscles in the body. Most of these special muscles are attached to your bones and enable you to move about. When you want to pick up a book, ride a bike, or take a bite of food, special nerves send a message to your brain, and your brain tells your muscles to get to work. Your brain gives very specific instructions to your muscles, saying which muscles should contract and pull your bones here and there. And that's exactly what they do – they pull on your bones. Your muscles never push, they only pull. When they pull, they tighten up and get shorter, but fatter. If you've ever watched a worm move, you can imagine this action happening. In the same way a worm shrinks down and gets shorter and fatter before it propels itself forward, so your muscles shrink down and get fatter when they contract.

Skeletal muscles are also called **voluntary muscles** because you can control them by thinking about controlling them. This means that you can think to yourself, "Self, lift your right arm up into the air and wave good-bye to your cousin." Immediately, your right arm will lift up and begin waving. You can volunteer to move these muscles, or you can volunteer not to move them. Voluntary muscles are, well, voluntary!

Although they are voluntary muscles, skeletal muscles can also respond with **reflexive** (rih flek' siv) action. In other words, your muscles can move immediately because of something that has happened, without any thought coming from your brain. Here's an example: once something bit my husband, and he automatically flung his arm towards the biter to ward it off – without even realizing or thinking about the fact that the biter was our toddler! No one got hurt, but it shows how a reflex action happens without your thinking about it.

Has a doctor ever tapped below your knee with a little hammer? What happened? I bet the lower part of your leg popped up! That's another example of a reflexive action. Doctors do this test to see how well your nervous system is working, and it is called "testing your reflexes." Essentially, it tells them how well some of your nerves are working. You can test your reflexes now!

When a doctor hits your knee with a hammer, your leg pops up without your thinking about it, because it is a reflexive action.

Try This!

Sit in a chair and cross one leg over the other (at the thigh level). Now, tap firmly on the crossed leg about a centimeter below the kneecap. Did your leg jerk outwards? How far out? This knee-jerk reaction is your body's reflex, or involuntary movement.

Your skeletal muscles also use reflexive movement when you get startled. When something surprises you, you might jump or tighten up very quickly. This happens because your muscles are often partially contracted, and that little scare causes them to jump into action. If you are very relaxed, your startled response is much less extreme. This partial contraction is called **muscle tone**. Everyone has muscle tone. Without it, you would collapse on the floor. You see, your muscles are always partly contracted because of signals that the brain sends them without your thinking about it. Only when you are sleeping do your muscles become so relaxed that they lose much of their tone.

So, while you're awake, your muscles are always working. Some people, however, are more toned than others, depending on how often they use their muscles. If you have a lot of muscle tone, your muscles are contracting quite a bit even when you are just sitting there. If you quit using your muscles by sitting all day, your muscles will lose some of their tone. Children tend to have more muscle tone than adults because they are more active. However, a stressed and nervous person will have more tone than normal, and will produce a greater startle reflex when someone says "Boo!" That's because the muscles were already partly contracted and ready to jump.

Tendons

So how do all your muscles move? Well, it's all in the way God designed them. Skeletal muscles attach to bones with tendons. Your tendons are like tough strings or cords that pull on the bone at specific spots when the muscles contract. Most skeletal muscles have at least two tendons, one at each end of the muscle. You can feel one of your tendons, your Achilles (uh kil' eez) tendon, where muscles of your calf attach to your heel bone. If you grab the back of your leg right above the heel, you will feel what seems to be a flexible, movable bone. That's actually your Achilles tendon. It was named after a person in Greek mythology by the name of Achilles. This is how the story goes: when Achilles was an infant, his mother dunked him in special water that would supposedly protect his whole body from harm. However, because she held him by the heel when she dunked him, the heel did not get wet and did not receive the water's special protection. So that heel was the only place where Achilles was vulnerable in war. Since this tendon attaches to the heel, we call it the Achilles tendon,

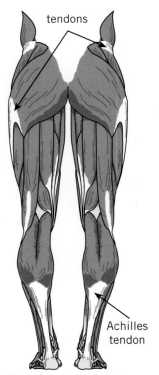

tendons

Achilles tendon

The white strips in this drawing are tendons, where the muscles attach to the bones.

after Achilles. It's actually the strongest tendon in your entire body. Tendons, as you may remember, are made from the fiber collagen. This makes them strong and flexible. Even though they are strong, if you overuse your tendons, they can get inflamed. We call that **tendonitis** (ten' duh nye' tis).

Did you know you can watch your tendons in action? It's true. Hold your hand in front of your face so you are looking at the back of it. Form your fingers into a claw. Now, move your pointer finger up and down. The thin line moving under your hand's skin is the tendon that connects one of your skeletal muscles to your pointer finger.

Moving Skeletons

In order to move your skeleton around, your muscles must contract and pull – remember, they cannot push. So, that same muscle that moved your bone up can't push the bone back down. When that muscle relaxes, the bone isn't pushed back into place. Another muscle (or just gravity) must pull the bone back into its original position.

Because of this, muscles must work together with other muscles. One muscle (or group of muscles) pulls the bone in a certain direction, and another muscle (or group of muscles) pulls the bone in the opposite direction. These pairs of muscle groups are called **antagonistic muscles**. When one set is contracting, the other is relaxing. For example, the bones in most joints have **extensor muscles**, which open the joint wider, and **flexor muscles**, which do the opposite and close the joint.

Let's study how this works by looking at your arm. Here you have an antagonistic pair of muscles: the **biceps** (bye' seps) **brachii** (bray' kee eye) and the **triceps** (try' seps) **brachii**. Look at the drawing on the left to see where these muscles are. The biceps muscle causes your elbow to bend or flex, so it is called the flexor muscle. The triceps muscle causes the elbow to straighten or extend, so it is called the extensor muscle. When the elbow bends, the biceps muscle tightens and the triceps relaxes. When the elbow straightens, the triceps muscle tightens and the biceps relaxes. Let's see how this works. With your palm up, stick your arm out straight. Put your other hand palm-down on the upper part of the arm that is stretched out, between the elbow and shoulder. Now, flex your forearm, like you are trying to make a muscle. Do you feel the bulge growing? That's your biceps brachii contracting to flex your forearm. In other words, when you make a muscle on your arm, the lump that is supposed to impress everyone is your biceps brachii contracted. Now, take the hand that is feeling your biceps and put it on the underside of your arm between your elbow and shoulder. Once again, your palm should be touching your arm. Now straighten your forearm so your arm is completely straight again. The muscle you feel now is your triceps brachii contracting to straighten out your arm.

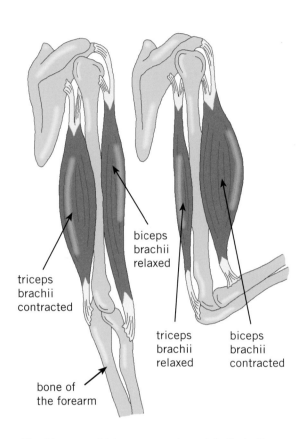

The biceps and triceps are antagonists that allow you to flex and straighten your forearm.

Although it would seem like the relaxing of the biceps would cause the forearm to straighten, it does not. The triceps must contract as well. It takes more effort to flex the forearm than it does to straighten it, because gravity pulls it down, which often helps straighten the forearm. Also, we tend to use our biceps more when we lift things. So usually your biceps are stronger than your triceps. Without use, a muscle will become weak and shrink. We call this condition **atrophy**. If you've ever seen someone's arm after a cast has been removed, you've seen atrophy. Atrophy is simply when some part of the body shrinks in size and becomes weak, usually because of disease, injury, or lack of use. Children use their muscles much more than adults because they run around and play, using a lot of energy. Most adults don't run and play as much as children do, so they usually have to make an extra effort to get the exercise they need to keep their muscles from atrophying.

Muscle Cells

So, what's going on inside your muscles? What kinds of cells make muscles work? Why do muscles get bigger when you use them a lot, and why do they shrink down when they aren't used much? Let's explore the inner workings of muscles to understand all of this.

Like all other living things, muscles are made up of cells. Skeletal muscle cells look different from your typical cells because they are long and thin – like a long piece of thread. In addition, a single muscle cell has lots of nuclei. Do you remember learning about the cell's nucleus? Well, "nuclei" is the plural of "nucleus." While most cells have only one nucleus, muscle cells have many. What's even more amazing is that each muscle cell is extremely thin, but it is long enough to run the entire length of each muscle. These long cells are bound together with all the other long cells to make the entire muscle. Because muscle cells are long and thin, they are often called **muscle fibers**. A muscle fiber is just another name for a skeletal muscle cell.

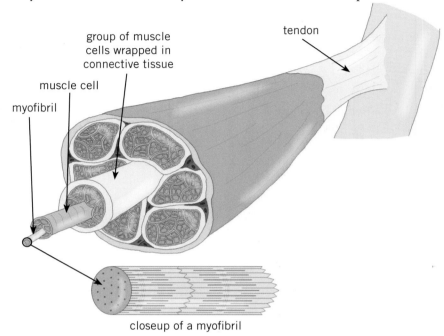

A muscle is composed of bundles that each contain many muscle cells. Each muscle cell contains many myofibrils, which contain filaments.

Inside each long, slender skeletal muscle cell are strands of protein called **myofibrils** (my' uh fye' brilz). Inside each myofibril are even smaller strips called **filaments** – some thick and others thin. They alternate: thick and thin, thick and thin. This makes the muscle look striped under a microscope. We call these stripes **striations** (stry aye' shunz). All skeletal muscles are **striated**.

Each muscle cell is wrapped in its own cell membrane, but it is also wrapped in connective tissue. Groups of muscle cells are wrapped together with more connective tissue. Do you remember learning about connective tissue in Lesson 1? It connects things. Muscle cells are "connected" by being bundled in connective tissue. The entire muscle is also wrapped with connective tissue. This connective tissue goes on to form the tendons at each end of the muscle. Do you remember that tendons are made of collagen? What does that tell you this connective tissue is made of? Collagen, of course! Because of this special system of collagen-wrapped muscle fibers and bundles of muscle fibers, each and every cell in the skeletal muscle is attached securely to the bone it will move. God created collagen to be quite useful, didn't He? It's used in our bones, tendons, and muscles.

Try This!

Let's take a close look at muscle fibers. We can do this by studying a piece of beef brisket. You see, beef is the muscle of a cow. We eat this cow muscle when we eat hamburgers and steak. The cow's muscle is very similar in appearance to a human muscle. So, let's take a close look at the muscle of a cow to see what we are talking about. Brisket is the best meat to study, because it has longer strips of muscle fibers. Often, the fibers have been cut in other types of meat, making the muscle fibers very short. However, if you can't find brisket, any piece of meat will do (except hamburger, which is ground-up muscle).

Begin with a lightly-cooked piece of meat, because it will make it easier to separate the fibers. Be sure to wear gloves for this activity. Using a toothpick, start separating the pieces of brisket into long, thin strips. What do you notice about the strips? Do you see how they are bound together? Can you snap the muscle fibers in half? Why not? Use a magnifying glass to study the muscle fibers up close. If you have a microscope, try to find the muscle cells.

Beef is cow muscle, so studying it helps us understand our own muscles.

Get a Move On

What is it that allows your muscles to move? Well, to move, your muscles need energy, which you get by eating good food and breathing oxygen. Your body breaks down your food, turning it into vitamins, minerals, and other things, including a sugar called **glucose** (gloo' kohs). Glucose and oxygen are used to make energy that can power your muscle cells. Without glucose, oxygen, vitamins, and minerals, your muscles would not work properly. But how do these things reach your muscles?

God designed a life-giving substance that flows through your body. This life-giving substance carries all that your body needs to do the things it needs to do. This special substance that carries all the nutrients you need throughout your body is called **blood**. Within every muscle are many blood vessels. Not only do the blood vessels bring in nutrients, they also carry away waste. Muscles, like everything else that is alive, make waste. One important waste product is carbon dioxide, also known as CO_2. That waste is picked up by your blood and carried to your lungs, where you actually breathe it out! We'll discuss this more in another lesson.

Even though your muscles need glucose, oxygen, vitamins, and minerals, they also need something to control them. Muscles are controlled by nerves. These nerves are connected to your brain through your spinal cord. They carry messages from your brain, telling your muscles to move. If a person's spinal cord has been damaged, he might not be able to move certain muscles, because the nerves in his spinal cord have been damaged. Without these nerves, muscles cannot receive messages from the brain telling them to move, and we say that the person is **paralyzed** in that region of his body.

God created an amazingly detailed plan when He designed your nerves and muscles. Depending on which muscle we are talking about, one nerve can control anywhere between ten to *more than a thousand muscle fibers* (remember, a muscle fiber is another name for a skeletal muscle cell). Because of this, you can trigger very small parts of certain muscles, like those in your hand – where very fine movements are needed in order to perform detailed activities like writing, sewing, playing the guitar, or performing surgery. In these muscles, where you are able to make delicate movements, you typically sacrifice strength in order to obtain precision. In the muscles where you have a lot of strength, you don't have as much control, nor as many different kinds of movements. For example, your leg muscles are very strong. Each nerve in your leg muscles stimulates a great many muscle cells, so your legs can generate a lot of force. Yet, your hand nerves can stimulate just a few muscle cells, allowing for a great many different kinds of small movements. This is the same with your face. The nerves in your face can stimulate just a few muscle cells, which allows for a huge number of different expressions on your face.

Let's Face It

Although many skeletal muscles attach to bone and create movement at the joints, not all skeletal muscles attach bone to bone with a joint in between. Many muscles in your face attach to bone and then to skin! This allows you to express yourself with those muscles. Interestingly, only people have such an enormous variety of facial expressions. These expressions give clues to a person's thoughts, feelings, and personality. Most animals can't make facial expressions. After all, have you ever seen a snake look surprised or a bird smile? A few animals, like monkeys and dogs, have a limited number of expressions they can make. The number of facial expressions you can make depends on how many muscles you have in your face, and human faces are packed with muscles. God intended this so we could communicate with one another simply by changing our expressions. Also remember that unlike the animals, we are created in the image of God. Since God communicates with us so effectively, it is not surprising that He made us so that we could communicate effectively with each other.

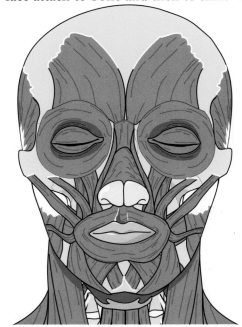

People are able to make lots of facial expressions because we have many muscles connecting the skin on the face to the bones of the skull.

Have you heard that you use more muscles to frown than you do to smile? That's not really true. You use roughly the same amount of facial muscles when smiling and frowning. However, scientific researchers have found that people actually feel happier when they smile – even if they have no reason to smile. They've also found that people feel sad when they frown, even if they have no reason to frown. Somehow, the expressions on your face can influence your emotions!

Try This!

Let's see if you can produce emotions in your body through facial expressions. Try to make your face reflect the feelings that go along with the following expressions: 1) sadness 2) anger 3) fright 4) loneliness 5) excitement 6) joy 7) contentment.

Which expressions made you feel better? I bet they were excitement, joy, and contentment. I guess that means that if you want a happier life, just put a smile on your face! After all, "*A cheerful heart is good medicine…*" (Proverbs 17:22)

Contracting Muscles

Because multiple nerves stimulate each muscle, you can actually contract a muscle with varying amounts of strength. You can also contract anywhere from just a small part of the muscle to nearly all of the muscle at one time. Think about using your biceps brachii muscle. You can contract your biceps just enough to move your forearm a tiny bit. This uses just a few of the nerve cells that control your biceps brachii. However, if you use a lot more of your nerve cells to contract your muscle more fully, you can lift a heavy rock. Also, God designed your muscle cells to "take turns" while your muscles are working, so that each individual cell does not get over-tired. That way, no single cell has to be working all the time. This very elegant system allows us to do many different activities with only a few different muscles.

Ballet dancers have strong muscles because they use them so much.

Let's experiment with our muscles. Get a timer and set it for five minutes. Next, hold your arm out straight out away from your body and keep it there. Do you remember which muscle is being used to extend your forearm so your arm is straight? It's your triceps brachii muscle. Muscles attached to your shoulder allow you to lift your arm so it is out and away from your body, so you're actually using several muscles right now as you hold out your arm. This is a pretty easy exercise, isn't it? Just holding your arm out to the side isn't really very hard. However, you are using muscles, and muscles do eventually get tired. See if you can hold your arm out for the full five minutes.

Your arm feels tired, doesn't it? What happened? Well, you simply exhausted the muscles you were using. While you were holding your arm up, the cells in the muscle were actually taking turns doing the work, but eventually, they became tired. As a result, it began to hurt to hold your arm up. The nerves didn't get tired, but your muscles did. If you did this activity every day, eventually the muscles you use would get stronger, increasing in size and increasing in the number of blood vessels bringing in nutrients and carrying out waste. This would result in your muscles being able to hold your arm out longer before becoming tired. Unless you engage in an activity like ballet, you aren't used to holding your arm out from your body for very long. A ballet dancer's regular practicing increases the strength in her muscles, allowing the dancer to easily hold her arm out for a long period of time. The more you use your muscles, the better they work!

Mighty Muscle Mitochondria

You've probably noticed that when you're running or exercising hard, you begin to breathe faster and harder. This is because of the microscopic mitochondria within each muscle cell! Do you remember what the mitochondria do for the cell? You probably remember that the mitochondria supply the cell with energy to do work. These mighty mitochondria use the glucose and oxygen brought by the blood to make energy so the muscles can move. Because muscles do a lot of work for the body, God created muscle cells with many mitochondria. When you're working your muscles hard, the cells need a lot of energy, so the mitochondria require more oxygen and glucose to create more energy.

Where do you think your blood gets all the oxygen? Well, take a deep breath. You just supplied your entire body with more oxygen. How the oxygen gets from your mouth into your blood is really cool, but we'll talk about that in a later lesson. For now, just know that when those muscles are working hard, your body says, "Breathe faster and harder because we're running out of oxygen!"

Sometimes, at the moment you need your muscles to work, you just can't give them the amount of oxygen they need. This usually happens when you are working your muscles a lot harder than you normally work them. When this happens, the cells switch to a different method of generating energy. This method doesn't require oxygen, so it is called **anaerobic** (an' uh roh' bik), which means "without oxygen." Unfortunately, cells cannot produce as much energy anaerobically, so this method is only used when oxygen is not getting to your cells quickly enough (like when you are in the middle of a long, hard run). Not only is anaerobic energy production not as good as **aerobic** energy production (energy production that uses oxygen), but it also creates a waste product in your cells called **lactic** (lak' tik) **acid**. Lactic acid can cause cramping during a workout, and it is part of what causes sore muscles after you exercise.

Have you ever noticed that once you have been running for a while, you continue to breathe hard even after you stop? If your muscles aren't moving so much anymore, why do you still need to breathe hard? Well, the main reason is the lactic acid I told you about. Once you get to the point where your muscle cells have to work anaerobically, lactic acid builds up. You have to get rid of it eventually, and your body can do that if it has oxygen. So to rid your cells of the lactic acid that built up while you were using your muscles a lot, your body will continue to require extra oxygen delivery and waste removal for some time after you finish working out.

Although working out is hard and can cause cramping

Muscles of the neck move the head and hold it up.

Muscles of the shoulders move the arms.

Muscles of the thigh move the leg below the knee.

Abdominal muscles move the torso and help with breathing hard.

Muscles of the calf pull the heel up.

Muscles in the shin region move the feet and toes.

Your muscles have been designed to work hard, and you benefit when you use them.

and soreness, it is very, very good for your muscles. God created our bodies to function best with plenty of physical exercise. In fact, He designed our bodies to work hard every day. It wasn't long ago that people had to walk everywhere they needed to go – including to the well to get water each day. People didn't always have cars to carry them places. They had to work hard just to eat. People grew their own food by tilling the soil and planting seeds; they ground their grain by hand and exercised their muscles just to make each meal. Everything required muscle: making clothes, building homes, cleaning the dishes. So our muscles have been designed to be used a lot.

You might remember what happens if muscles are not used. They begin to atrophy. But if they are used regularly, your muscles will become larger and stronger and will make your whole body healthier. You see, contracting your muscles frequently not only increases your strength, but it also keeps you well! People who exercise regularly have a much happier life than those who don't, and they are less likely to get many illnesses, including cancer, heart attacks, adult-onset diabetes, and depression. The drawing above points out some muscle groups and what they do when you exercise.

Growing Muscles

So how do your muscles get bigger? Believe it or not, it is not by producing more muscle cells. The number of muscle cells in a muscle stays the same once you have matured. Your muscles get bigger because you actually grow more myofibrils. It works sort of the same way as when your skin develops a callus. Have you ever seen someone with calluses on his hands? When a person uses his hands in a way that rubs the skin raw, the skin on that area grows thicker to protect the hands when they are being used. For example, a guitar player's fingertips become irritated, raw, and sore from pressing on the guitar strings. To compensate for this, the skin grows thicker on the tips of the fingers so that it does not hurt when he plays the guitar. Muscles grow in much the same way. When you use a muscle more than it is accustomed to being used, it might get sore at first. To compensate, the myofibrils in the muscle grow thicker. The more the muscle is used, the thicker the myofibrils grow so the muscle can handle the strain of whatever activity you are doing. The muscle may grow so big that it even bulges under the skin. If you increase the strain, it will (over time) grow bigger still. That is why someone who works out with weights has to continue increasing the amount of weight lifted in order to keep increasing the size of his muscles.

Incidentally, all that strain on the muscles also puts strain on the bones. So, the bones do the same thing that the muscles do – they compensate for the strain. People who exercise and work out regularly have stronger bones than those who do not. That means it's harder to break a fit person's bones. Very old people who do not work out at all often have thin, frail bones. When that occurs, a small fall can break their bones. It's important that, as you grow older, you continue to get a lot of exercise to keep your bones dense and strong.

Pack the Protein

Although it takes glucose, oxygen, vitamins, and minerals to make muscles move, your muscles need something else to continue growing. You see, muscle is made up almost entirely of protein. So you can help to keep the muscles God gave you healthy by eating plenty of protein. Eggs, dairy products, meat, fish, and many grains and vegetables contain protein that can be used to build your muscles.

Before you move on, explain in your own words what you have learned about muscles so far.

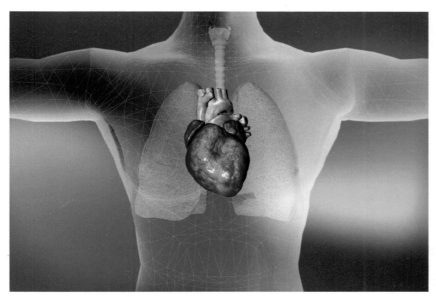

There is only one place you will find cardiac muscle – in your heart.

Cardiac Muscle

I've told you a lot about skeletal muscles, but you have other kinds of muscle in your body. As you already learned, you have cardiac muscle as well. Cardiac refers to the heart, and that's exactly where you'll find cardiac muscle. In fact, that's the only place you'll find it. Cardiac muscle contracts and relaxes, contracts and relaxes, contracts and relaxes without your thinking about it. In other words, it's involuntary. You can hear the phases of this contraction and relaxation – it's called your heartbeat, and it happens quite a lot. In fact, your heart beats about 100,000 times per day!

If you want a healthy heart, you need to exercise it – just like your other muscles. How do you do that? Well, you should do some kind of exercise that raises your heart rate, making it beat faster and harder, for at least thirty minutes. Decide on what kinds of exercise you can do that will make your heart beat hard for thirty minutes. You should do these exercises at least three times a week or more.

Smooth Muscles

We are going to learn all about your heart in another lesson, so for the rest of this lesson, I want you to learn about your smooth muscles. Remember how skeletal muscles have stripes, which we call striations, when looked at under a microscope? Well, smooth muscles are described as "smooth" because they do not have striations. Interestingly enough, cardiac muscle has some striations, but not as many as skeletal muscle.

Skeletal Muscle

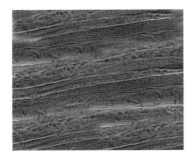

Cardiac Muscle

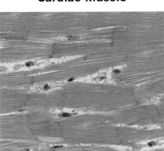

Smooth Muscle

When viewed under a microscope, skeletal muscle (left) has definite striations (the many vertical lines in the picture). Smooth muscle (right) has no striations, and cardiac muscle (center) is somewhere in between.

Smooth muscles are found in places like your blood vessels, stomach, intestines, and bladder. These special muscles are involuntary. You don't have to think to make them work. They just do their job automatically. What do they do? Well, the smooth muscles that operate your intestines move food that comes from your stomach the rest of the way through your digestive system and out the other end. The way they do this is similar to how you might squeeze the last bit of toothpaste out of a tube. The muscles either contract or relax, making the size of the intestine tube smaller behind where the food is. That pushes the food farther along down the intestine.

Try This!

To get an idea of how your smooth muscles push food through your intestines, get a nylon stocking (one that nobody will use anymore) and a ball of clay that is bigger around than the stocking is wide. Cut the top off the nylon stocking. Place the ball of clay inside and begin working it through the stocking. Consider the different ways you have to squeeze the clay to get it through the stocking. Imagine the muscles of the intestines relaxing and contracting to get a ball of food through. That's how those smooth muscles work!

Your intestines move food like you have to move a ball through a stocking.

What Do You Remember?

What are the three kinds of muscle tissue in your body? Which of those muscle kinds are voluntary, and which are involuntary? What is muscle tone? What are tendons? Where is your Achilles tendon? What are antagonistic muscles? What do muscle cells have a lot of that give them energy? What substances help muscles move? What do muscles need to grow? How do you keep your cardiac muscle strong? Name two places in your body where you find smooth muscles.

Personal Person Project

For this lesson, you will add muscles to your Personal Person. If you have the *Anatomy Notebooking Journal*, there are muscles for you to cut out and add to your Personal Person. If you don't have the *Anatomy Notebooking Journal*, draw what you see in this picture or muscles like those shown in the drawing on page 64.

Notebooking Activity

For this lesson's notebook activity, you will create the Muscle Times – a newspaper that teaches people all about muscles. Include the information you learned in this lesson. You can draw pictures of muscles or print some from the Internet. It would be a good idea to add information about how to help your muscles get stronger and work better. You could even include different ideas for getting exercise. Be creative and have fun!

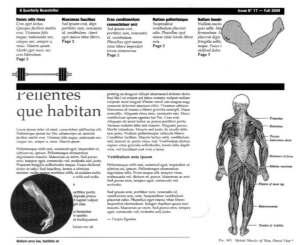

Experiment
Growing Muscle

As we discussed, the more you use a muscle, the bigger it grows and the better it works. We are going to do an experiment over the next two weeks to build some of the muscles in your hand.

You will need:

- A clothes pin that opens when you squeeze on it and closes when you release
- A timer
- Graph paper (Only older students need this.)
- A pencil
- Paper

1. Predict how many times you can open and close the clothespin with your right hand in one minute.
2. Set the timer for one minute and see how many times you can actually do it.
3. Were some of your muscles tired at the end of the minute? Do you feel any soreness? Where?
4. Write down how many times you could open and close the clothespin in one minute.
5. Repeat steps 2-4 with your left hand. Write down the results.
6. Now you are going to time how long it takes for you to fill one sheet of paper by writing your first and last name over and over again. Begin the timer, and when the paper is completely filled, stop the timer. Write down how long it took you to fill the page. Do you think you used any of the same muscles for the writing activity that you used with the clothespin activity?
7. What do you think will happen if you did steps 2-4 every other day for two weeks? Use a Scientific

Speculation Sheet to write down your hypothesis (best guess).

8. Choose your dominant hand (the hand with which you write) and do steps 2-4 with that hand every other day for two weeks. This means you will do the clothespin exercise seven more times with your dominant hand.

9. You should now have eight measurements of how many times you could open and close the clothespin in one minute. Do you notice anything about the numbers? Do they get bigger or smaller as the days go on?

10. Do the writing exercise again. Were you able to fill the page faster this time?

Most likely, you saw that as you continued to do the exercise, you could open and close the clothespin more times. This is because your muscles got stronger by doing the exercise over and over, and they were able to go longer without getting tired. Since you use some of the same hand muscles to write, you probably were able to fill up the page with your name faster at the end of the experiment than you were able to at the beginning of the experiment.

Older students: Make a graph of the results. The graph will have the day you did the experiment on the horizontal axis and the number of times you could open and close the clothespin on the vertical axis. It should look something like the graph below.

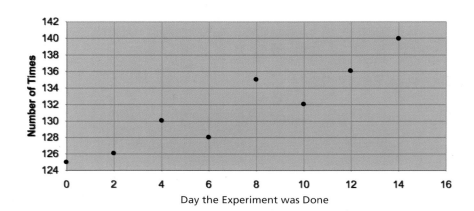

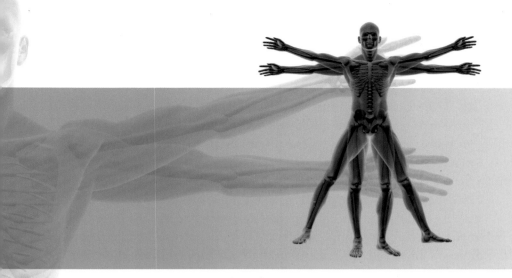

the DIGESTIVE AND RENAL SYSTEMS

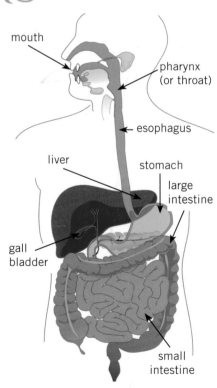

All your life you've been eating food – taking it in your mouth and swallowing it. You've been doing this every day since the day you were born. You've probably never gone a single day without food or liquid. So, have you ever wondered what happens to the food once it goes down your throat? Have you ever wondered why you need to eat, or what difference it makes whether you eat cake or vegetables, candy or steak? Why does your mother insist you eat certain foods over others – especially since the foods she wants you to eat don't always taste as good as the foods you want to eat? And why does the waste product that comes out the other end of you not look anything like the food that sat on your plate? What is going on here? I'll tell you; it's called **digestion** (dih jes' chun). You see, your **digestive system** actually takes the food you eat and converts it into materials your body needs to live, repair itself, and grow.

The part of your digestive system that the food passes through is sometimes called your **gastrointestinal** (gas' troh in tes' tuh nul) **tract**, or GI tract for short. It is also called the **alimentary** (al' uh men' tuh ree) **canal**. We'll use all these terms throughout this lesson. Just remember, your digestive system is made up of all organs involved in how your body takes in food, uses what it can, and eliminates the rest. On the other hand, the alimentary canal, gastrointestinal tract, and digestive tract are different names for the same thing – the part of the digestive system through which food actually travels.

Your digestive system includes all organs related to taking in food, breaking it down, making it useful to the body, and getting rid of what is left over. It contains some organs (like the liver and gall bladder) through which food never passes.

69

Down the Hatch

So, what happens when you bite into a cheesy piece of pizza? Well, you're eating all sorts of nutrients with that one bite. For right now, I want to concentrate on the three that you need the most of: **carbohydrates** (kar' boh hi' draytz), **proteins** (pro' teenz) and **fats**. We'll study other nutrients (like vitamins) in more detail in the next lesson. Just remember that these are the three main nutrients your body needs, and your digestive system handles each of these types of nutrients in different ways. So, in goes the pizza. Now what?

The main thing your body is interested in doing with that pizza is breaking the huge carbohydrates, proteins, and fats into smaller chemicals that can be used by your body. You know, when your body has the right amount of carbohydrates, proteins, and fats, it can use them to do all sorts of things, like building, growing, and repairing cells, tissues, and organs. All the rest of the stuff you eat goes through your body and is tossed out the other end through a process called **defecation** (def' ih kay' shun). Defecation is the last step in digestion where your body removes the waste products from the food you have eaten – it then gets flushed down the toilet. So, how does that pizza get from your plate into your cells and out the other end? Let's find out!

Grand Opening Mouth

I bet you think digestion begins once you swallow your food, but that's not true. You see, digestion begins the minute the food enters your mouth. Your mouth is well equipped to begin this process. After all, it's equipped with the masseter muscles (do you remember what those are?) and has no problem biting, tearing, and grinding your food. This is the mechanical part of digestion. **Mechanical digestion** is the grinding up and moving along of the food. This grinding begins at the top – with those beautiful pearly white teeth God gave you.

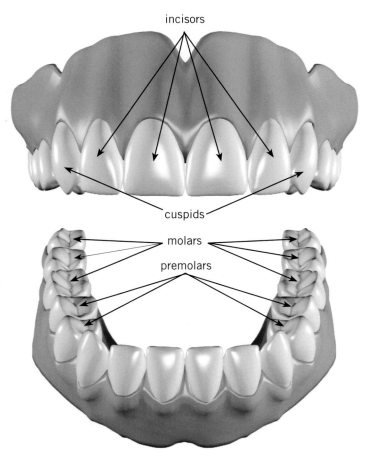

incisors

cuspids

molars

premolars

Terrific Teeth

Use a mirror to look in your mouth. Notice the different shapes of your teeth. Each tooth has a different purpose. Your **incisors** (in size' orz) are sharp and useful for biting. Can you find your sharp incisor teeth? Your **cuspids** (kus' pidz), sometimes called your canines, are also sharp and are made for tearing food. Those flat **premolars** (pre moh' lurz) and **molars** (moh' lurz) in the back of your mouth are great for grinding up your food. If you had only tearing teeth, you would have a hard time chewing your food, wouldn't you? You would have to swallow big chunks of food, like an alligator or a shark does. If molars were the only teeth you had, you would only be able to eat food that was already in small pieces, ready to grind. God designed your teeth just perfectly for your **diet** (which is just another word for the things you eat).

Look at the diagram of a tooth on the next page. Your teeth are coated with an extremely hard, white, shiny substance called **enamel** (ih nam' uhl). The enamel is not a living substance, but it is the hardest substance in your body – harder even than compact bone! The **dentin** (den' tin) under the enamel is alive, because it contains cells. It supports

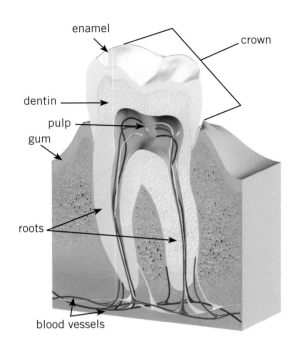

enamel

crown

dentin

pulp

gum

roots

blood vessels

A single tooth is amazingly complex!

the enamel and absorbs shocks that could otherwise damage the tooth. Below the dentin is the **pulp**, which is where the nerves and blood vessels can be found. Have you ever felt pain in your teeth when you were eating ice cream or drinking hot cocoa? That's the pulp shouting, because it's sensitive to hot and cold.

Look at your teeth in the mirror again. The **crown** of the tooth is the part you see above your gums. It's covered with enamel. Your teeth are firmly set into your jaw bone by their roots.

It's really important to take care of your teeth. You see, there are bacteria waddling around in your mouth right at this very moment – lots of them. These bacteria are really hungry. Do you realize that you feed the bacteria every time you eat and don't brush your teeth? You see, whenever you fail to brush your teeth well, you give the bacteria a feast – they feed on the food left on your teeth. The more you feed them, the more bacteria you'll have living in your mouth. As if that's not gross enough, after these bacteria eat the leftover food, they produce waste. Yes, they do. Not only does this waste stink, causing bad breath, it is acidic and breaks down the enamel covering your teeth. Then you can get holes, or cavities, in your teeth. **Ouch!** If a cavity reaches the pulp, it can hurt really badly! You could even lose your tooth. You don't want to lose your teeth because they are necessary for mechanical digestion – they are good grinding, chewing, and tearing machines. If that doesn't make you want to brush your teeth three times a day, I don't know what will!

Try This!

Let's find out what acid can do to your teeth. Find an old baby tooth and some soda pop. If you can't find a human tooth, see if you can find an animal's tooth. Put the tooth in a glass and then pour a lot of soda pop in the glass. Every few days, remove the tooth and examine it, then put it back in the glass with some more soda. You'll be amazed at what happens to the tooth. You see, soda pop contains acid. Just like the acid that the bacteria in your mouth make, this acid can eat through the enamel of the tooth. It takes some time to eat through the enamel, because the enamel is hard. Eventually, however, it does happen.

Super Saliva

So, the pizza goes in and you start to chew it, which begins the process of mechanical digestion. This just breaks the food down into little bits. At the same time, however, **chemical digestion** begins. You see, the food is made up of all sorts of chemicals, and those chemicals need to be changed into smaller chemical components so your body can use them. That's what chemical digestion does. To get this started, your mouth makes **saliva** (suh lye' vuh), otherwise known as spit.

If you ever thought your spit wasn't important, think again! God put a lot of thought into making spit just right. It is 99% water, but it has special chemicals that protect your teeth, break down food, and defend your mouth against infection. Believe it or not, saliva also makes your food taste better. Only God could design a substance that does so many things all at once.

If you put your tongue on the inside of your cheek across from your second upper molar, you'll detect a

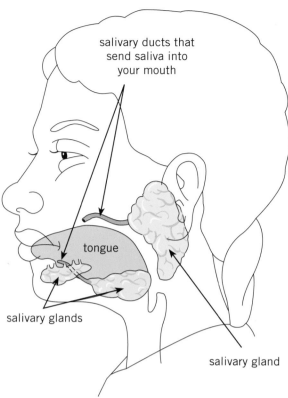

The salivary glands around your mouth produce saliva, which then goes through salivary ducts and into your mouth.

little bump on your cheek. That's the opening of one of your salivary ducts, which takes saliva from your salivary glands to your mouth! You just felt one duct, but as the drawing to the left shows, there are other salivary ducts that come from other salivary glands in your chin.

So how does saliva help with chemical digestion? Well, remember those special chemicals it has? One of those special chemicals is an **enzyme** (en' zyme). Enzymes are chemicals that help your body do all sorts of things, including breaking your food down into smaller chemical components. The enzyme in your saliva is **amylase** (am' uh layz'), and it breaks down starches. Starches are found in bread, potatoes, rice, beans, corn, and many other foods.

Try This!

Let's watch amylase in action. You will need a saltine cracker and a mirror. Put a one-inch piece of the saltine cracker on your tongue, but do not chew it. Leave it there on your tongue and watch what happens. Do you notice any special taste? Try putting a one-inch piece of cheese on your tongue. Did the same thing occur?

Did you see the cracker break down? Did you taste something sweet while that was happening? You should have, because the amylase in your saliva breaks down starch (which was in the cracker) into sugar. However, there is no starch in cheese, so nothing unusual should have happened when you put the cheese on your tongue. This process of breaking down starch is the beginnings of chemical digestion. Pretty cool, huh?

Another interesting thing about salivary glands is that they actually begin working even before you get the food in your mouth. In fact, when you see, smell, or sometimes even think about food, your brain tells the salivary glands to start making saliva. Even more astonishing is that without the moisture of the saliva, your taste buds don't even work at all. We'll explore your taste buds more in a later lesson. For now, be thankful God provided you with all that spit in your mouth!

Terrific Tongue

Once when I was a child, I remember opening the refrigerator and gazing at an enormous tongue sitting in there. Yes. A tongue. It was huge. It was a cow's tongue. My parents actually intended for us to eat that thing! Thankfully, they accepted my refusal, and I can tell you today that I've never eaten a tongue. However, I've accidentally bitten my own tongue. Have you ever done that? Well, the tongue is another great invention of God and is an extremely important muscle in the digestion process. Do you think it's a voluntary

Someone might actually eat this cow's tongue one day!

or an involuntary muscle? Well, you can control it, so it is a voluntary muscle. Since it's a voluntary muscle, you should work on controlling your tongue, because it can get you into trouble. In fact, controlling your tongue should be a priority in your life, as God tells us it's one of the marks of spiritual maturity. In James 3:2, He says, *"If anyone does not stumble in what he says, he is a perfect man, able to bridle the whole body as well."*

So what's the tongue up to now with our piece of pizza? Well, as you chew, your tongue makes special movements that form the food into a little ball that we call a **bolus** (boh' lus). You don't really have to think about those movements. They just happen automatically as you chew. Then, after you swallow the bolus, something amazing happens – the food goes down, down, down your esophagus right into your stomach, and not into your lungs!

Why is this amazing? Well, think about what happens when you breathe in – air passes through your nose or mouth, down your throat, and into your lungs. The **larynx** (layr' inks), which leads to the lungs, is right in front of the entrance to the esophagus, which leads to your stomach. How does air know to go into your lungs and food know to go into your stomach? Though the larynx and esophagus are close to one another, God has designed a clever system that prevents you from inhaling your food. Let's see how it works!

Gently feel the roof of your mouth. It is called your palate. You'll notice that the front part is hard, while farther back it's soft. When you swallow, the soft palate moves up in a way that actually seals your nasal cavity, so you can't breathe and swallow at the same time. If you ever laughed while you were drinking and had the drink come out your nose, it was because laughing broke that seal. That's why it's important to swallow before you talk, laugh, or breathe. While you're swallowing your food, a special flap of cartilage called the **epiglottis**, which is in the back of your throat, drops down over the larynx, which leads to your lungs. So, the pizza bolus you just swallowed is guided safely over the epiglottis and into the esophagus. Look at the diagram below to see how this works.

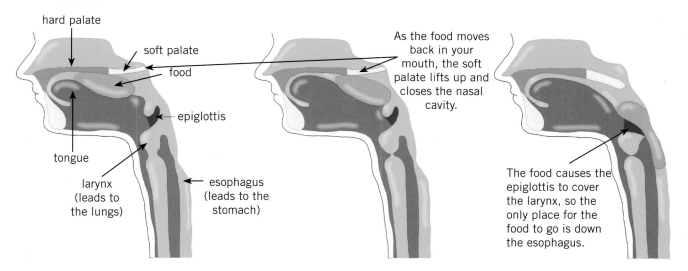

hard palate
soft palate
food
epiglottis
tongue
larynx (leads to the lungs)
esophagus (leads to the stomach)

As the food moves back in your mouth, the soft palate lifts up and closes the nasal cavity.

The food causes the epiglottis to cover the larynx, so the only place for the food to go is down the esophagus.

If by chance some food does touch the pathway leading to the lungs, God designed an emergency response – your body coughs vigorously to clear the airway. Pretty amazing, huh?

Before moving down into the stomach, explain what you have learned so far.

Stirring Stomach

So down your food goes – down your esophagus on its way to your stomach. But what if you are upside down? Will your food still go into your stomach? Indeed, food does not *fall* into the stomach, it is *pushed* into your stomach by the muscles of the esophagus. So, even if you're hanging like a monkey upside down, you can still eat. I don't recommend it, however. How does the esophagus push food down to the stomach? Well, it has two layers of muscle. One layer has long muscles that run the length of the esophagus. The other layer is circular and wraps around the esophagus. These muscles contract and relax in a wave down the alimentary canal. They push the food through, much like the ball you pushed through the nylon in the last lesson.

At the meeting point between the esophagus and the stomach there is a special circular muscle called a **sphincter** (sfingk' tur) that opens and closes by contracting and relaxing. This special sphincter muscle is called the **gastroesophageal** (gas' troh ih sof' uh jee' uhl) **sphincter**. Study that word a bit and see if any parts of it look familiar. (You should always do this when you are studying big words in science, because a lot of times they are made from many smaller words that help you figure out the meaning.) In this case, "gastro" means "stomach." You probably recognized that the "esophag" part refers to the esophagus. So, basically the gastroesophageal sphincter is the stomach/esophagus sphincter. Most of the time, the gastroesophageal sphincter muscle is contracted (closed) so that food being digested doesn't exit the stomach and go back up the esophagus. When the muscles of the esophagus push the bolus along, the sphincter relaxes so that food can pass into the stomach.

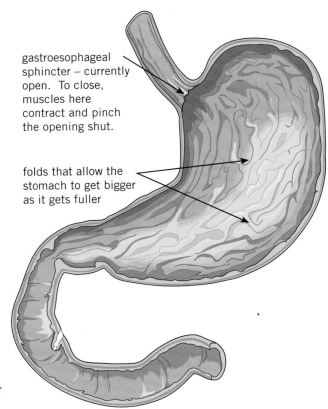

gastroesophageal sphincter – currently open. To close, muscles here contract and pinch the opening shut.

folds that allow the stomach to get bigger as it gets fuller

This is a drawing of what an empty stomach would look like if you cut it in half and looked inside.

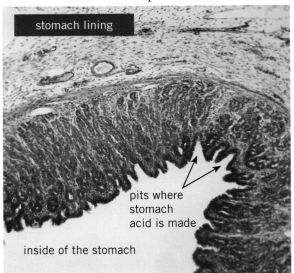

stomach lining

pits where stomach acid is made

inside of the stomach

This image (magnified about 200x) shows your stomach lining is composed of many cells, including cells that make stomach acid and cells that protect the tissue from that acid.

The stomach is a muscular pouch designed to store food for a short time. When it is empty, it has lots of folds, like a wrinkled-up paper sack. When you eat, the folds flatten out, increasing the size of the stomach and allowing it to hold more food, about 1.5 liters worth. The big plastic soda pop bottles at the store are 2-liter bottles, so your stomach can hold almost as much as one of those bottles. Once expanded, your stomach is the widest part of your digestive tract. However, once your food leaves, your stomach folds back up again.

Now, here's where things get really wild for that piece of pizza. The smooth muscles in the stomach begin churning, while acids begin burning, and your pizza is turning – and possibly yearning for this storming, stirring stomach to stop! The acid in your stomach does two things. First, it kills many different kinds of bacteria. God knew that occasionally we would eat bacteria which aren't good for us. So He designed this acid as a powerful protection against infection. The second thing acid does is break down the proteins you eat into smaller chemicals that your body can absorb and use. Do you remember which parts of your body need protein? I'll tell you – all your cells, but particularly your muscle cells.

You should know that your stomach acid is potent and powerful. It's so strong that it would burn your skin – and even your internal organs quite severely – if it came in direct contact with them. Yet God, in His very great wisdom, provided you protection from this acid. He designed special cells that produce a thick layer of **mucus** (myoo' kus) to line the stomach. Mucus is a thick, slimy liquid (like snot) that keeps the stomach acid from coming into contact with the stomach wall. That way, the acid can be powerful enough to digest your food, but it won't harm your stomach!

Have you ever heard of acid indigestion? Well, that happens when acid escapes the stomach and gets up into the esophagus. If this happens, it can hurt so much that people describe it as their heart burning, or heartburn, which makes sense because the esophagus isn't too far from the heart. In fact, some people have thought they were having a heart attack, when they were really experiencing this terrible pain from stomach acid in their esophagus. Thankfully, the gastroesophageal sphincter usually does a great job of keeping the acid in the stomach and out of the esophagus.

Creation Confirmation

Do you remember what an enzyme is? It is a kind of chemical that helps the body do a lot of things, including chemically breaking down the food you eat. Well, your stomach has a powerful enzyme in it called **pepsin** (pep' sun). It is needed to break down the proteins you eat into smaller chemical components that your body can use. Cells that line your stomach need to produce pepsin so your stomach can do its job. However, your cells cannot really produce pepsin, because *they are made of proteins*. If a cell made pepsin, it would destroy itself! Nevertheless, your stomach needs pepsin to do its job, so your cells have to make it. Guess how they do it. They make a slightly different, inactive chemical that *turns into pepsin when it is mixed with stomach acid*. That way, they don't harm themselves, but they supply the stomach with what it needs. Think about that for a moment. Your stomach acid is necessary for digestion, but it can harm the stomach. Your body knows this, so some of the cells in your stomach produce mucus to protect it from the acid. In addition, your stomach needs pepsin, but that can harm the very cells that make it. However, your body knows this, so your cells make a slightly different chemical that doesn't become pepsin until it leaves the cells. Isn't that amazing? It shows that your body is the result of careful forethought and design!

So, your stomach acid is important. However, what's even more amazing is that God designed this acid to begin flowing in your stomach the moment you took that first bite of pizza! Because it's such powerful stuff, your stomach tries not to have much acid flowing around, except when you're eating.

Stomach Stories

Since air is often swallowed along with food and liquid, the gastroesophageal sphincter occasionally relaxes to let the air out of your stomach. This results in a **burp**, or a belch. Sometimes the body detects something bad during digestion. When this happens, your body goes into red alert and everything works in reverse. Violent contractions push food up the esophagus and out the mouth. We call this lovely experience **vomiting**. Can you now understand why your throat burns when you vomit? Your stomach pushes acid out, along with the offending food, and your esophagus and throat get burned by the acid.

Has your stomach ever growled at you? It sometimes sounds like it is an angry animal, doesn't it? Well, those rumblings from the stomach are the result of the stomach muscles contracting around air caught in its folds when it is empty, or at least nearly empty. The air vibrates in response to the muscle contractions, and you hear those vibrations as your stomach "growling."

Try This!

You are going to simulate your stomach in action! You will need two Ziploc bags, a piece of bread, and water. Put the bread in a Ziploc bag and pour some water inside so it covers the bread. Now seal the bag and place it inside another bag for protection. Begin squeezing and squishing the bag as your stomach might do. What happens to the bread? This is what happens to your food when your stomach churns it.

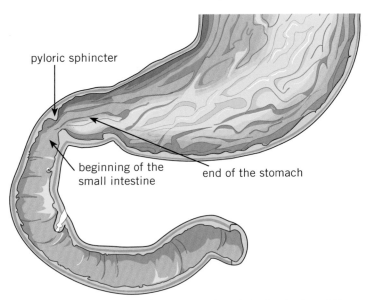

pyloric sphincter

beginning of the small intestine

end of the stomach

This shows what the end of the stomach and the beginning of the small intestine would look like if they were cut in half and opened.

Chyme to Go

After your pizza is thoroughly mixed up and some of the protein in the crust and cheese is broken down, the pizza becomes a thick liquid we call chyme. It takes about 3 or 4 hours for food to become chyme. This chyme is then slowly released into the small intestine. This is accomplished by a different sphincter, the **pyloric** (py lor' ik) **sphincter**. The pyloric sphincter works carefully so that only a small amount of chyme goes into the small intestine at a time. That way, the food can be slowly processed to get all the nutrients out before sending it along. It's here, in the small intestine, where most of the chemical digestion occurs. Proteins, carbohydrates, and fats are broken down by enzymes into small chemicals your cells can use. Finally, these tiny chemicals are absorbed through the wall of the small intestine and into the bloodstream. It takes about

3 hours for your food to be processed by the small intestine. That's pretty fast, considering your small intestine would be about 20 feet long if it were stretched out in a straight line!

Try This!

To get an idea of how long your small intestine is, get a measuring tape and mark off 20 feet somewhere outdoors or in a long hallway. It's amazing something that long can fit inside your little body!

You may wonder why the small intestine is called "small," since it's so very long. Well, it's because it is small in diameter – about one inch wide, while the large intestine is much wider. The small intestine has three parts: the **duodenum** (doo uh dee' num), the **jejunum** (jeh jyou' num) and the **ileum** (ill' ee uhm). Even though the duodenum is only the first foot of the 20-feet-long small intestine, the greatest amount of chemical digestion occurs there. The jejunum is the next eight feet, and the last 11 feet are the ileum. These two parts of the intestine are mostly responsible for absorbing the nutrients produced by chemical digestion into the bloodstream. Hopefully you are choosing to eat food that contains lots of good nutrients!

Like the rest of the digestive tract, food moves through the small intestine because smooth muscles that line the small intestine push it along. God designed the small intestine very specifically – with the ability to send proteins, carbohydrates, fats, vitamins, and minerals to the body. He covered the entire surface of the small intestine with **villi**. What are villi? They are tiny projections, like little fingers, that make the inside wall of the intestine look almost like it is lined with velvet. Cells inside the villi transport the nutrients from the small intestine to blood vessels that are scattered throughout the villi.

Why does your small intestine have villi? Well, in order to absorb the nutrients from digestion, the cells need to *come into contact with the nutrients*. If the wall of the small intestine were smooth, the nutrients would have to brush up against the wall to come into contact with a cell. The villi point down into the intestine, exposing more cells to the nutrients that come from digestion. This allows for more contact between the cells and the chemicals they are supposed to absorb. Without these villi, your small intestine would need to be *over two miles long* in order for the same amount of nutrients to be absorbed by the cells.

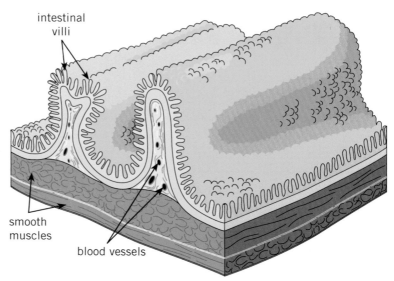

This is a drawing of the wall of the small intestine.

Take a moment to tell someone what you learned about the stomach and the small intestine.

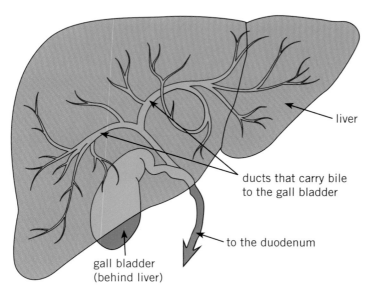

Your liver has many functions, but one that is directly related to digestion is the making of bile, which is stored in the gall bladder.

Living Liver

Once the nutrients are absorbed into the bloodstream, they pass through the **liver**. The liver has many important roles in the body. In the liver, nutrients are processed and then either stored or distributed throughout the body. The main thing the liver stores is glucose. This is the little sugar molecule your cells (especially your brain cells) rely on for energy. If you have a lot of glucose in your body, your liver strings the little glucose molecules together to make **glycogen** (gly′ kuh jen). When your body runs low on glucose (like between meals), the liver breaks the glycogen back down into many glucose molecules and sends them into your bloodstream. The liver also stores various vitamins and minerals, including iron.

The liver is like a huge chemical factory. One of the chemicals it makes is called **bile**, and it is a very important chemical for the digestive process. After being made in the liver, it is delivered to the gall bladder so it can be stored and concentrated. As the chyme goes into the small intestine, the **gall bladder** then releases concentrated bile into the duodenum. What does bile do for you? Well, bile helps break down fats! Fats don't mix well with water, so in a watery mixture like chyme, they form large clumps. Bile breaks those large clumps into smaller clumps that are easier for your small intestine to digest. You can do the same thing!

Try This!

Pour some cooking oil (which is just a kind of fat) into a bowl of water. Notice that there are large globs of oil floating on the water. Now, use a fork to whisk the oil a bit in the bowl. Do you see that the drops of oil are smaller? Now add some liquid dishwashing soap and whisk the oil, water, and soap together. Let the bowl sit for a while and come back every now and then to check on it. Once all the bubbles are gone, notice the difference. You should see a lot of very tiny drops of oil now, because the soap acted like bile, breaking the fat into tiny clumps that are much smaller than they were before.

The liver is very valuable to your survival. It does so much for you. Not only does it make bile and store nutrients like glucose, vitamins, and minerals, it actually takes chemicals that would gradually build up and poison you and makes them less harmful to you. For example, when your body breaks down proteins, one of the products is ammonia. Your body needs to break down proteins, but the ammonia that is made by this process is very bad for you. If it were to build up in your body, it could end up killing you! To keep this from happening, your liver takes the ammonia and uses carbon dioxide to convert it into a substance called **urea** (yoo re' uh). That substance is then sent to your renal system so it can be removed from your body. We'll discuss the renal system at the end of this lesson.

Pancreas Potential

Another organ used for digestion is your **pancreas** (pan' kree us). It produces hormones as well as digestive juices. One hormone it makes is insulin, which controls how much sugar is in your blood (we call that your blood sugar level). We'll discuss insulin in a later lesson. The pancreas also produces **pancreatic juice**, which is secreted into the small intestine when chyme arrives. This juice has two main functions. It neutralizes the acid in the chyme, making it far less harsh, and it contains important enzymes that help to digest fats and break down proteins and starches. Since much of what you eat (meats, fruits, vegetables) contains cells, you need to be able to break down the contents of these cells. Without a pancreas, your body would not be able to break down the food you eat and give the nutrients to the rest of your body. Aren't you glad God gave you a pancreas?

Now you've learned all about the first part of digestion. Explain what each organ you've learned about does for your body. Don't forget the teeth, saliva, stomach, small intestine, liver, and pancreas.

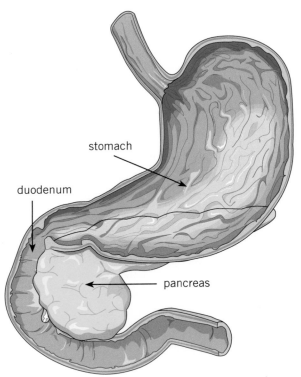

One function of the pancreas is to inject pancreatic juice into the duodenum.

Large Intestine

So now your piece of pizza is about to leave the small intestine, but it doesn't look anything like that cheesy slice you put in your mouth about six hours ago. At the meeting point between the small and large intestines you find the **ileocecal** (ill ee oh see' kul) **sphincter**, which opens and closes to let the liquid chyme from the small intestine enter the large intestine.

The large intestine has three parts: the **cecum** (see' kum), the **colon** (kol' un), and the **rectum** (rek'

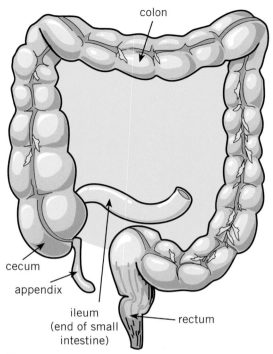

colon

cecum

appendix

ileum
(end of small
intestine)

rectum

Your large intestine receives chyme from your small intestine and pulls out water and salts to put into your body.

tum). Its main job is to form the chyme into **feces** (fee' seez). Feces are the waste material that we evacuate into the toilet. In order to make feces, the large intestine removes the water from the chyme. This water is absorbed through the wall of the large intestine into the blood stream. Almost five cups of water (1 liter) are sent from the large intestine into the bloodstream every day. Now you would think that water from your large intestine would contaminate your body, but when the large intestine pulls it from the chyme, it only pulls clean water and some salts, both of which are good for your body. So basically, the large intestine is drying up your food so you can use the water that was in it!

Here is the interesting part. You see, living in the large intestine are millions upon millions of friendly bacteria. Now remember, some of these microscopic creatures cause illness, but not all of them! You have a **symbiotic** (sym bye' ah tik) relationship with some of them. If you studied zoology, you may remember what symbiotic relationships are. They are relationships where two living things work together to benefit one another. Have you ever hosted guests at your house? Well, you are hosting the friendly bacteria in your colon right now. Say hello to your guests! You both benefit from their presence in your colon. You benefit because these friendly bacteria digest some otherwise indigestible things in the chyme, and in doing so, they produce vitamins that are absorbed by your body.

The most significant vitamin these bacteria produce is vitamin K. Without vitamin K, you would bleed to death when you cut yourself. Vitamin K helps your blood to clot. Vitamin K is also present in some foods you eat, such as broccoli and spinach. The intestinal bacteria synthesize some B vitamins as well. Not only do you benefit from this relationship, but the bacteria benefit as well, because they have a place to live with lots of food. You're a good host! Interestingly, these friendly bacteria remain friendly only if they remain in the colon. If they travel elsewhere in the body, they can cause serious illness. Fortunately, this is rare.

The wormlike tube on the cecum is called the **appendix**. For a long time, scientists didn't know what its role was in the body, and some actually thought it had none. Of course, anyone who understands that we are made by God knows that He wouldn't make something in our body that was useless! Thus, creation scientists always thought it had a job in the body. Well, in 2007 at least one of its jobs was discovered. Remember those friendly bacteria? The appendix provides a safe place for them to hide when your body gets an illness that could kill them. You see, some diseases you get are bad for your friendly bacteria. So they hide out in the appendix until the illness is gone, and then they go back to living in your intestine and making vitamins for you! This allows you to recover from those kinds of illnesses much more quickly.

Did you ever wonder why beans can cause gas to form in your intestines? It's because beans contain complex carbohydrates, which are very resistant to being broken down by your digestive system. The bacteria in your intestines can digest these carbohydrates, but in doing so they release several different gasses, which can cause some rather embarrassing side effects. We call this side effect **flatulence** (flach' uh lentz). Some people are better at digesting beans than others. I'll say no more on this subject!

As wavelike muscle movements continue to move the chyme through the large intestine, more and more water is removed. This can take anywhere from half a day to three days, depending on your diet. A diet high in **fiber** (plant material that you can't digest) speeds up the process. Since the chyme moves more quickly through the large intestine, less water is removed from it, so it stays reasonably soft. This means that a diet high in fiber keeps you from getting constipated. Also, some research shows that colon cancer could be caused by food staying in the colon too long, so eat your fiber!

The rectum is the last segment of the colon. It is here that the feces are stored before being eliminated from your body. The last sphincter muscle is found here. It's one you can control – which gives you time to find a restroom when you need one.

The Renal System

In addition to your feces, your body also produces another kind of waste – the liquid form, which we call urine. Urine is produced in your **renal system** by special organs called **kidneys**. Put both of your hands behind your back just below the back of your rib cage. That's where your kidneys are located! The kidneys are a pair of bean-shaped organs. In fact, kidney beans got their name because they look like kidneys! Hmm, which came first, kidneys or kidney beans? Probably the beans, since God created plants before man. But, that's another discussion.

So, what's going on with the kidneys and the renal system? Well, your body produces a lot of waste while it's doing all the things God created it to do. Much of this waste enters the bloodstream where it is sent to the kidneys for processing. The kidneys are your body's water control and blood-cleaning organs. When your blood enters your kidneys, it encounters millions of tiny purifying systems that remove wastes and water. The "waste water" is what we call urine. Urine trickles down through tubes called the **ureters** (yoo ree' turz). You have a ureter attached to each kidney. These ureters lead down to a pouch in the very middle of your pelvic girdle. This pouch is your **bladder**. When your bladder is full, it sends a message to your brain: it's time to find a restroom!

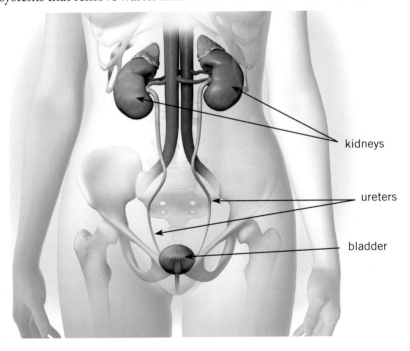

kidneys

ureters

bladder

The amount of urine your kidneys release depends on many things, including how much water you drink and how many salts you are eating. Salts are essential for your body. Table salt (the stuff you sprinkle on your food) is just one kind of salt. There are many, many more in creation, and if your body doesn't have the right kinds of salts in the correct amounts, it can stop operating properly. In fact, if you don't have the right salt levels in your body, it doesn't digest food properly, you lose muscle tone, and you can lose the ability to think properly. If you drink a lot of water, the salts in your body get diluted, which upsets the balance of salts in your blood. To bring your body back into balance, your kidneys get rid of the extra water. That's why you need to go to the restroom often when you drink a lot of water. That water is being pulled from your blood by the kidneys so your salt levels go back to normal!

Now if you don't drink very much and you sweat a lot, you lose water from your body. When this happens, your kidneys quit taking water from your blood, and you don't urinate very much. Interestingly enough, when you sweat a lot, you also lose salts, because your sweat contains a lot of salt. People who live and work in the desert, for example, must eat salty foods to keep the salt content of their bodies normal. This is also the reason athletes drink sports drinks instead of water. The sports drinks have salts in them, and that restores salts to their bodies.

Another important function of your kidneys is to rid your body of urea. You learned earlier that your body needs to break down proteins during digestion, but that process makes ammonia, which is bad for you.

These paper cups once held sports drinks that marathon runners used to replenish their water and salts. When you run and sweat for a long time, you lose too much water and salt, which must be replenished. You can see from the number of cups that a lot of runners needed water and salts!

The liver takes that ammonia and turns it into urea, and your kidneys put the urea in your urine. In fact, urea is what gives urine its characteristic odor.

Our kidneys are just another sensational creation of God. All your blood flows through at least one of your two fist-sized kidneys every five minutes! More than 280 times each day, all the blood in your body gets a thorough cleaning! Isn't that amazing?

What Do You Remember?

What is the white outer layer of your tooth called? What is the layer right below that called? What is the hardest substance in your body? Name a few things saliva does for you. What is the name of the pipe that food goes down after you swallow it? How do your stomach and esophagus keep from getting burned by your own stomach acid? What is the food called when it enters the small intestine? What happens in the small intestine? Which organ is like a huge chemical factory? What do the kidneys do?

Notebooking Activities

In addition to writing down all the fascinating facts you learned about the digestive and renal systems, you will create a comic strip of the digestion process. A comic strip is simply a series of boxes with different actions happening in each box. You will animate, or bring to life, a piece of food. You can choose any food you wish. Follow it as it goes from your plate and through your digestive system. Imagine that the food has feelings. How does it react to each of the stages of digestion? Express all that in your comic strip.

Personal Person Project

Today, you will add the renal system and the digestive system to your Personal Person. If you have the *Anatomy Notebooking Journal* that goes with this course, there are drawings in the appendix for you to cut out and use. If not, use the drawings on pages 69 and 80 as well as the pictures below as a guide for drawing them yourself.

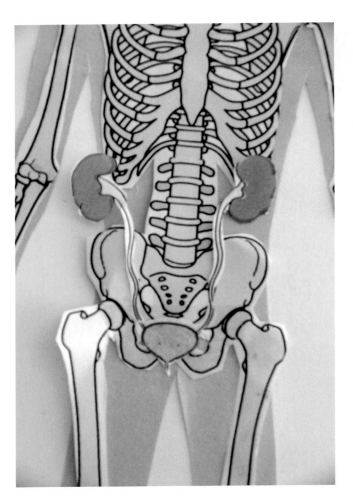
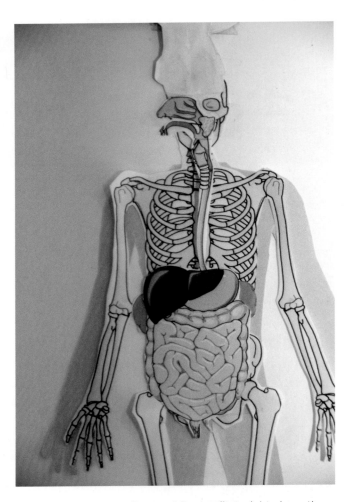

Since your kidneys are behind most of your digestive system, put the renal system on your Personal Person first, right above the skeletal system. Put the digestive system on top of that, right under the muscular system.

Project
Design a Digestion Theme Park

Have you ever been to a theme park? It's a large amusement park with rides, shows, and other attractions. If you haven't been to one, let me explain how theme parks work. A theme park is based on a theme. For example, Disney World's Animal Kingdom is based on an animal theme. There are roller coasters, spinning rides, shows, parades, restaurants, and shops – and they all have to do with the animals of the world. If you go to the course website I told you about in the introduction to this book, you will find a link to Disney World's Animal Kingdom theme park as well as links to a few other theme parks.

Your project is to create a theme park based on the digestive system. You will design rides and attractions that explore the different organs and processes of digestion. Follow these steps to create your theme park:

1. On a piece of paper, begin writing down all the different organs of the digestive system that you want to highlight in your theme park.
2. Write down or draw out ideas for rides, experiences, restaurants, and other attractions. This is called brainstorming. Make sure everything you include teaches or demonstrates the different steps of digestion.
3. Use a separate sheet of paper for each attraction and begin designing what each will look like and how it will function. You should plan to use illustrations and labels, as well as explanations.
4. On a sheet of scratch paper, design the layout for your theme park and create a map. You can go to the Disney site or other sites linked from the course website to look at their maps for ideas. You will need an entrance, restrooms, plants, signs, restaurants, and food stands.
5. Using a large sheet of construction paper, begin work on the final layout and design. You will make your final work three-dimensional by drawing each attraction and folding a small tab on the bottom to stand it up on your paper. You can use tape or glue to affix each attraction to your park.
6. Be sure to take a friend or family member on a "tour" of your Digestion Theme Park!

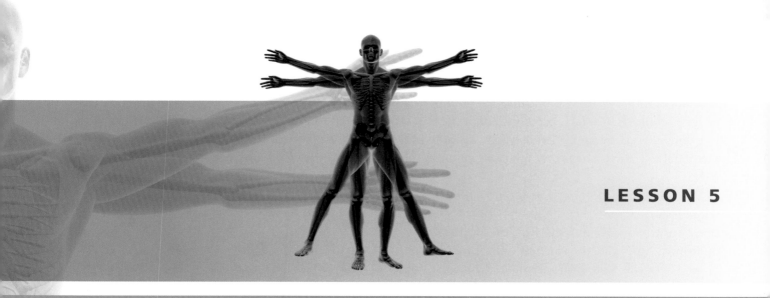

HEALTH AND NUTRITION

You've probably heard the words, "Eat your vegetables" more than once in your young life. Many of them are green; they sit on your plate looking up at you. What's so special about them? How do they benefit your health any more than the big chocolate cake you ate at your birthday party? That's what we are going to explore in this lesson.

Nutrition is the science that explores how your body uses the food you eat. In this lesson, you'll come to understand the importance of fruits and vegetables in your diet as well as why you need protein, sugar (yes, I said sugar), and fat (yes, I said fat). You will learn different words for "sugar" and "protein" that you'll use throughout the rest of this course. And the best part is that

The science of nutrition tells you why the apple is better for your body than the candy bar.

this information will be an important part of your life forever and ever. No matter where you go, what you do, or how long you live, you'll always be eating. And as long as you're eating, you'll want to know what your body needs to stay healthy and active – so you can fulfill God's great plan for your life!

We'll go over the different things your body needs to survive. Then, you'll do fun activities to remember what you've learned. Are you ready for a nutritious adventure? Let's go!

Necessary Nutrients

The foods and liquids that God made for you to eat and drink have **nutrients** (noo' tree uhntz), which are the substances found in food and drink that your body needs to be healthy, work, and grow. The main nutrients are water, carbohydrates, fats, proteins, vitamins, and minerals. You need vitamins and minerals in tiny amounts, but you need large amounts of water, carbohydrates, fats, and proteins. Your body can store most of these nutrients so that you have them in times of need. Do you remember which organ stores a lot of nutrients? Your liver!

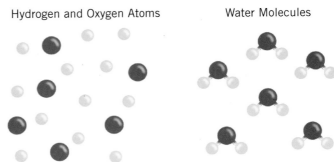

Hydrogen and Oxygen Atoms   Water Molecules

Here is something you should know before we get started: all the things you put into your body are chemicals, and all chemicals are made up of **atoms**. Atoms are the smallest unit of a chemical. Some chemicals (like iron) are made from just one kind of atom. However, most of the chemicals used by your body are made of many different kinds of atoms linked together to form molecules. For example, one atom of oxygen and two atoms of hydrogen make a very important substance we call water. Chemists actually call water "H_2O" because of this. The "H" represents hydrogen atoms and the "O" represents oxygen atoms. If you add another

Atoms join together to become new chemicals called molecules. The yellow balls are hydrogen atoms, and the red balls are oxygen atoms. As atoms (left), they are colorless gases. When two hydrogen atoms link to one oxygen atom (right), they make water.

atom of oxygen to this molecule, it is no longer water. Instead, it changes the chemical completely, turning water into hydrogen peroxide (H_2O_2), which is that stuff that bubbles when you put it on a cut. Since I've mentioned water, let's learn more about that important nutrient.

Most experts say you should drink eight glasses of water every day.

Win with Water

What do you suppose is your body's most important nutrient? Well, did you know that an adult can go without food for a month or more? Yet, he could not do without water for more than a few days. Most people would die after three or four days with no water. Even though it is that important to your body, water does not contain anything that your body can use to make energy. So, why do we need so much of it so often?

Well, it's because your body is about 60% water. Remove the water, and you're more than half gone. All your cells depend on water. Water is a part of your blood, which spreads oxygen and nutrients around your body. Also, water is used in the fluid that cushions your joints so you don't get injured as easily. Many of the chemical reactions that take place in your body require water. Water does a lot of things for you. Since you're made of water, and since your body is using it all the time, it's really important to get a lot of it.

So how much water should you drink? Many people would say to drink eight 8-ounce glasses of water a day. Although you get a lot of water from the food you eat (especially if you eat fruit), eight glasses isn't a bad idea. However, the best way to know if you're getting enough water is whether or not you feel thirsty.

When your kidneys determine that you don't have enough water in your blood, they send out a distress signal which says you are **dehydrated** (dee' hi dray' ted). This means your body is low on water. Your brain then triggers your thirst, and you find you want some water right away! Another way to tell if you are dehydrated is to look at your urine. If it's dark yellow, your kidneys are telling you to drink more water.

The reason you must continually get more water in your body is because you lose water all day long when you go to the restroom, sweat, and even when you breathe (the air you breathe out has water vapor in it). So you need to replace that water often. When you exercise, when it's hot outside, or when you're sick, you lose even more water!

Try This!

Are you surprised that you breathe water out of your body? Show yourself that it's true. Go up to a cool window and put your face really close to it. Now open your mouth wide and expel a huge breath from your lungs onto the window. What happens? You should have seen the window fog up. Why? Because of the water vapor in your breath. The water vapor hit the cool glass and formed water droplets, which is what you call "fog" on the glass. It's similar to how a bathroom mirror will fog up when someone is taking a shower.

This mirror is fogged up because water vapor from a hot shower hit it, and the coolness of the mirror caused the water to form tiny droplets on it.

Carb Control

Carbohydrates (kar boh hi' draytz), sometimes called "carbs" for short, are very important for your body. In fact, you should be getting most of your energy from the carbohydrates you eat. Have you ever had a gadget that ran on batteries, and as the batteries wore out, the gadget slowed down and eventually just stopped? What happened when you added new batteries? Your gadget popped back into action again! Well, carbs are like the batteries that make you go. Essentially, when carbs enter your body they turn into sugar, which is a source of energy.

Study the word carbohydrate for a moment. Do you see anything familiar in that word? The "carb" part refers to carbon, and the "hydrate" part refers to water, which is made of hydrogen atoms and oxygen atoms. That should tell you carbs are made from carbon atoms, hydrogen atoms, and oxygen atoms. You get most of your carbs from eating plants and plant products, like fruits, grains, and vegetables.

Simply Energetic

Simple carbohydrates usually taste mighty sweet. You can find them in many foods, including fruits, honey, and sugar cane (the plant from which we get white table sugar). We call them "simple" carbohydrates because, as carbohydrates go, they are pretty small molecules. That means your body doesn't take a long time to digest, process, and turn them into energy. This may seem like a great thing, but there are some problems with this kind of "quick energy."

You might remember how you felt after drinking a sugary beverage or eating candy when you were a toddler. Suddenly, you were bouncing off the walls – bursting with noise, excitement, and energy! Your mom may have even commented on how you had "overdosed" on sugar. Well, that's because some simple carbohydrates do that to you. They give you quick energy. But this energy doesn't last long, and soon after, you can find yourself feeling tired. That's part of the problem with eating too many simple carbohydrates: after spending a few hours super charged, you crash and burn, feeling tired, grumpy, and sad the rest of the day. Also, when your blood sugar falls after rising so fast, your brain thinks it's time to eat again. This can lead to eating too much food as your body tries to keep your blood sugar levels steady.

Simple carbohydrates are found in sweet things like table sugar, maple syrup, honey, and fruits. Now while the first three things on that list can result in problems like I just talked about if you eat too much

of them, fruits typically don't give you a "sugar rush." That's because there actually aren't a lot of simple carbohydrates in most fruits. The simple carbohydrate in most fruits is called **fructose** (frook' tohs). It is a very sweet sugar, so it doesn't take much to make a fruit sweet. Another reason fruits don't usually give you a sugar rush is that most contain a lot of fiber. The fiber helps to slow down how quickly the fructose is absorbed into the bloodstream by the intestines.

Unlike candy and other sweets that typically contain table sugar, fruits are also rich in vitamins, minerals, and other important nutrients. As a result, fruits are a healthy source of simple carbohydrates, while most sweets are not.

While fruits and table sugar both contain simple carbohydrates, fruits are a better source when it comes to your health.

Beans and nuts contain complex carbohydrates and fiber, providing a low glycemic index.

Complex Carbohydrates

The molecules that make up **complex carbohydrates** are larger than those that make up simple carbohydrates. Because of this, many complex carbohydrates take longer to break down in your digestive system, so they don't release energy into your bloodstream very quickly. Scientists have a measure for how quickly food releases energy into your bloodstream. It is called the **glycemic** (gly see' mik) **index**. The higher the glycemic index of the food, the more quickly you can get energy from it. Foods that have a high glycemic index might contain simple carbohydrates, or they might contain complex carbohydrates that your body can break down quickly. Potatoes, for example, have mostly complex carbohydrates in them, but your body breaks them down quickly, so potatoes have a high glycemic index.

If you eat a lot of food with a high glycemic index, you get a super energy kick, but later on you feel grumpy and tired. If you eat foods with a low glycemic index, the energy is released into your bloodstream more slowly. Beans, peas, and nuts typically have a low glycemic index.

Carbohydrates: The Inside Story

Do you remember the cracker from the previous lesson – the one that started getting digested on your tongue? It contained starch, which is a complex carbohydrate. As I told you then, an enzyme in your saliva was responsible for starting the digestion. It was actually taking a complex carbohydrate (called starch) and breaking it down into simple carbohydrates.

Inside your body, all complex carbohydrates must be broken down into simple carbohydrates, and even some simple carbohydrates like **sucrose** (table sugar) have to be broken down even further. Eventually, your body likes to turn all carbohydrates into one very simple carb, called **glucose** (gloo' kohs). The more complex the carbohydrate, the longer it takes to be changed into glucose. This means that a meal of complex carbohydrates generally takes longer to digest.

Glucose is like a battery that keeps you running. It travels in your blood (you'll learn all about blood pretty soon). Because your blood carries the glucose around your body, we call the glucose in your bloodstream blood sugar. When the blood sugar reaches the cell that needs it, it is taken up by that cell and converted into energy. Do you remember where in the cell this happens? If you guessed "the mitochondria" you are correct!

But what happens when you eat more carbohydrates than you need? What do you think your body does with the extra? Well, God created your body to store glucose so you can use it later if you need it before you have a chance to eat again. Your liver changes the glucose into a complex carbohydrate called **glycogen** (gly' kuh jen), because glycogen is a more efficient way to store energy than keeping individual glucose molecules around. That way, if you get up early and go outside to play before eating breakfast, your body uses the glycogen, breaking it back down into glucose and sending it to the cells that need energy.

Now what happens if your body decides you have plenty of glucose *and* plenty of glycogen but then you eat more carbohydrates? If that happens, the glucose is changed into fat. You see, just as glycogen is a more efficient way to store energy than glucose, fat is an even more efficient way to store energy. If you have plenty of glucose and plenty of glycogen, your body decides to put the food into "long-term storage," which is fat. The only way to get rid of those fat stores is to eat less food and do an activity that uses lots of energy. That way, after your body uses up all the stored glycogen, it will start to use the fat stores, converting them into energy. So although eating too much fat can make you gain fat, eating too many carbohydrates can also make you gain fat.

Try This!

You can test the starch content of foods with iodine, so let's test various food items for starch! Begin with some iodine, a dropper, and several items of food. Cut open a few fruits and vegetables, including a white potato and a sweet potato. Now guess which items will have starch and which will not, based on what you have learned. Test the foods for starch by placing one drop of iodine on each food item. If the food item has starch, a chemical reaction will occur turning the place where the iodine was dropped dark blue, almost black.

Before you move on to proteins, explain what you have learned about water and carbohydrates.

Power-Packed Protein

Every cell in your body makes and uses **protein**. What is protein? Well, remember how we talked about molecules? Molecules are atoms that are linked together to make something completely new. Well, some molecules are called **amino** (uh mee' noh) **acids**. If you string those molecules together, you get a protein. So, if you want to sound smart, you can ask your dad to pass the protein instead of asking for the meat. But if you want to sound really, really smart, you can ask him to pass the amino acids. Of course, you might not get your food since he may not know what you're talking about!

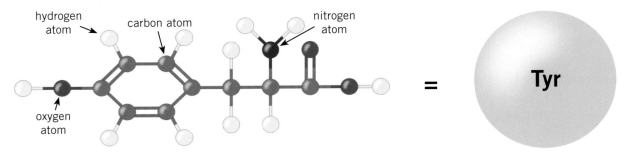

This is an amino acid: a molecule made up of many atoms linked together. This specific amino acid is tyrosine (ty' ruh zeen').

We will abbreviate the molecule as a single ball with three letters representing its name.

Proteins are made up of hundreds, sometimes thousands, of amino acids strung together. Most proteins have many of the same amino acids, but in different combinations and patterns. Now here's the amazing part – your cells make the proteins they need. Why do we need to eat protein if our cells can make the proteins they need? Because your cells need those amino acids to make proteins. Your cells need 20 different kinds of amino acids, and they can actually make most of them if they need to. But there are nine amino acids that your cells can't make. In order to get those amino acids, you must eat food that has proteins that contain them. These nine amino acids are essential for your cells to make the proteins they need, so we call these acids **essential amino acids**.

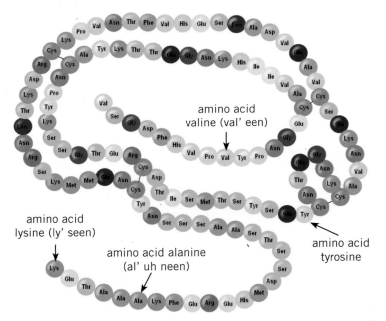

Many different amino acids linked up in a specific order make a protein. This particular protein is a simple one called ribonuclease (rye boh new' klee ays).

Getting the Essentials

Some foods have proteins that include all nine essential amino acids, while the proteins in some foods have only a few of them. If food has proteins that contain all nine essential amino acids, we say it is a source of **complete protein**. If it has only some of those amino acids, we say it is a source of **incomplete protein**. Complete proteins are found in eggs, milk products, and meat (such as beef, chicken, pork, and fish). If you eat some of these products each day, you'll provide your body with all the amino acids it needs.

Incomplete proteins are found in plant products. Beans, nuts, and grains all have different combinations of the nine essential amino acids in them, but none have all nine together. If you want to get all nine from one food, it must be a food that contains complete proteins. Vegans (people who do not eat any animal products) must eat different kinds of plant products to get all nine essential amino acids. For example, beans and rice together contain all nine essential amino acids, as does peanut butter on whole-wheat bread. However, you can't get all nine essential amino acids by eating whole-wheat bread and rice.

Proteins: The Inside Story

Inside your body, the proteins (remember – they are long chains of amino acids) are broken down into separate amino acids, which are then sent in the blood throughout the body to be delivered to the cells that need them. Unlike carbohydrates and fats, however, your body can't store proteins. If you aren't eating enough protein, your body will start breaking down muscle and other tissues in order to get the amino acids needed to keep you going. Obviously, this is only a short-term solution. I mean, you don't want to get rid of those strong leg muscles! It's better to get too much protein than not enough. If you eat more than your body can use, the extra amino acids can be used for fuel or changed into other useful molecules. You need a daily supply of amino acids in your diet if you want your body to be strong and healthy.

Try This!

Let's do some math to figure out how much protein you need each day. You need about 0.05 ounces of protein for every 2 pounds you weigh. To figure out how much you need, divide your weight (you measured that in Lesson 3) by 2. Then, multiply that number by 0.05. So, if you weigh 60 pounds, you need about 1.5 ounces of protein a day. You can get that much protein by eating four eggs or two small hamburgers. You see, it's easy to get the amount of protein you need – it just takes a little planning!

Eggs are an excellent source of protein.

The Skinny on Fats

If there's anything people try to avoid eating, it's fat. What they don't realize is how great fat is for them (in the right amount, of course)! If you were to cut out all the fat from your diet, you would get very sick indeed. Fat is used by the body every day to carry out some extremely important tasks. For example, fat stores important vitamins in your body. Fat keeps your skin healthy. It coats important cells so they can work properly. Some fats help your body heal. Your body stores energy as fat to be used when you can't eat.

Fat also contributes a great deal to the taste and texture of food, and it is the fat in a meal that gives you the feeling you've had a satisfying amount to eat. That's why people who eat a low fat diet often feel hungrier than those who do not – and they often end up eating more as a result. Scientists call fats **lipids**. When people go to the doctor, they often get their blood checked to see what kinds of lipids are floating around in there.

Fundamental Fatty Acids

Fats are made up of molecules we call fatty acids. In fact, a fat molecule is three fatty acid molecules linked together with one **glycerol** (glis' uh rahl) **molecule**. So, we call fats **triglycerides** (try glis' uh rydz'). Your body makes the triglycerides it needs, but it can only make them if it has the right fatty acids. Well, just like there are some amino acids your body can't make, there are some fatty acids your body can't make. We call them **essential**

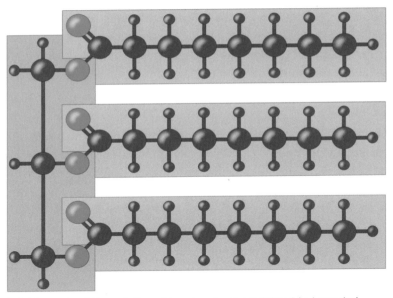

A fat is made of three fatty acid molecules (highlighted in lavender) linked together with a glycerol molecule (highlighted in pink).

fatty acids, and in order to remain healthy, you need to eat them so your body can make the fats it needs to make.

There are two essential fatty acids, called **omega 3** and **omega 6** fatty acids. Your body needs both kinds, but it needs more omega 6 fatty acids than it needs omega 3 fatty acids. That's good, because omega 6 fatty acids are easy to get. They are found in most of the oils we use in cooking. In fact, since it is so easy to get omega 6 fatty acids in your diet, most people in the United States eat too many omega 6 fatty acids. Like most things your body needs, too many omega 6 fatty acids can be as bad as too few. The omega 3 fatty acids are harder to find in a typical diet. However, you can find them in salmon, tuna, sardines, dark green leafy vegetables, avocados, sesame seeds, walnuts, pumpkin seeds, and flax seeds.

Now remember, fats are made from three fatty acids linked by a glycerol molecule. So in order to give your body fatty acids, you must eat fats. There are two basic kinds of fats: **saturated fats** and **unsaturated fats**. Saturated fats are usually solid at room temperature. Butter and cheese, for example, are made of saturated fats. Unsaturated fats are usually liquid at room temperature. We normally call these fats "oils." Most cooking oils are made from unsaturated fats. Unsaturated fats are usually better for you, but saturated fats taste better to most people. Because of this, companies often take unsaturated fats and chemically convert them into saturated fats through a process called **hydrogenation** (hi' druh juh nay' shun). These saturated fats are called **trans fats** and are now known to be unhealthy. As a result, most food labels list the amount of trans fat in a food. Foods high in trans fats include fried foods (like French fries and doughnuts) as well as stick margarines.

Try This!

The grease stains on this bag tell you that it holds food with a lot of fat.

Let's see if we can determine how much fat is in the foods you have around the house. Pick a few different kinds of foods that you would like to test for fat content. You might choose cheese, fruits, cereal, peanut butter, cookies, condiments, or any other food you are wondering about. After choosing your samples, cut them into small pieces (about the size of a marble, for example). To compare fat content, all the samples need to be the same size. If the food can't be cut, simply try to match the amount to the rest of the foods.

Cut a paper bag into 2-inch square pieces, one piece for each food sample. Make a hypothesis about which foods will have the most fat and which may not have any fat at all. Then place each food sample on its own square and label that square with the name of the food. Rub each sample on its square for 20 seconds, and then let the samples sit on their squares for two hours. This allows time for the fats to be absorbed by the paper squares. After two hours, remove the samples. Using a hair dryer, blow dry each piece of paper. After the papers are dry, look for grease spots. Compare the sizes of the grease spots. Can you tell which foods contain the most fat?

Counting Calories

So how much fat do you need each day? The answer will require a little math. Are you ready? Many nutrition experts think that you shouldn't get more than 30% of your daily food calories (kal' uh reez) from fat. But, wait! What is a calorie?

I'm sure you've heard the word "calorie" coming from some adults now and then. It is a unit of measurement, like "pound" is a unit of weight and "inch" is a unit of length. The calorie is one unit we use to measure energy. To be precise, a calorie is how much energy it takes to heat 1 gram of water by 1 degree Celsius. That may seem a bit strange, but you get the picture – it's a way to measure energy. Now let's translate that to food. The number of calories in food tells you how much energy the food will give your body. Do you use a lot of energy? If so, you'll need to eat a lot of calories.

Now if you consume more energy, or calories, than you actually use, you'll probably get a little on the heavy side, because your body stores that extra energy as fat until you need it. So basically, you need a certain number of calories each day, depending on your size and activity level. If you sit around reading all day and never go out to play, you'll need fewer calories than active children. Either way, you should only get about 30% of your calories from fat.

This label tells you that of the 180 calories in a serving of this soup, 60 come from fat. To calculate the percentage, take the calories from fat, divide by the total calories, and multiply by 100: 60 ÷ 180 x 100 = 33.3%

Ounce for ounce, or gram for gram, fats contain a lot more calories than do proteins and carbohydrates. That means if you put the same weight of fat, protein, and carbohydrates on a plate, the fat would contain the most calories. So it doesn't take a lot of fat for you to get the fat calories you need. But how can you tell how much fat you're eating? One way is to look at food labels. They give a lot of information about how many calories come from carbohydrates, how many come from proteins, and how many come from fats. In fact, why don't you study some labels now to see what's in the food you have in your pantry!

Try This!

Let's take another trip to your pantry. Pull out some boxes, jars, and bags of food that have labels. Don't look too closely at the labels just yet. Based on what you've learned about nutrition, try to guess whether carbohydrates, proteins, or fats will have the biggest number on the label.

Older students: Use the math explained in the figure caption to calculate the percentage of calories that come from fat.

Explain what you have learned so far before moving on to the next section.

Victorious Vitamins

Have you ever taken vitamin pills? Does your mom buy you the delicious gummy vitamins, or does she get you the crunchy kind in colorful animal shapes? Maybe you take capsule-type vitamins that are harder to swallow. You may be wondering what all this vitamin business is about. You're told you need them, but why? Do you really need all those vitamins? Did you ever think that since two are good for you, twelve would be even better? Well, now it's time to explore these nutrients so we can answer such questions for you.

Although your body definitely needs vitamins, it only needs a very small amount to survive. If you don't get enough, you can get very sick. If you get way too many, bad things can happen to you as well. Thankfully, the food sources God made – plants and animals – usually have a great supply of all the vitamins you need. But sometimes these original vitamins get removed when the food is being prepared, especially when it is being prepared for storage. For example, white bread and whole-wheat bread are both made from the same wheat grains, but they are different because of the parts they use. The very outside part of the wheat grain is called the bran, and the inside part is called the germ. There's also a part in between the bran and the germ, called the **endosperm** (en' duh spurm), which is the white, starchy, potato-like part. It contains a lot of carbohydrates, but not many vitamins. God designed the grain to be a complex food with fiber, vitamins,

Whole-wheat bread is made with the entire grain of wheat, which contains important vitamins your body needs to make you strong, healthy and smart. White bread is made by removing the vitamin-rich parts of the grain; so it has very little nutrition.

and carbohydrates. Most of the vitamins in the wheat grain are found in the bran and germ parts. However, people found out that if you remove the bran and the germ from the wheat grain, the potato-like endosperm will store for a long, long time without going bad. So some bread companies remove the bran and germ, which are where a lot of the vitamins are stored. What is left is simply the simple carbohydrate part of the bread, and that is used to make white bread. So it's healthier to eat whole-wheat bread instead of white bread, because whole-wheat bread uses all parts of the wheat grain, including those that have most of the vitamins.

Sometimes companies will "fortify" or "build up" foods with vitamins to make them healthier for you. For example, milk companies add vitamin D to their milk when they process it. This makes the milk healthier, because vitamin D helps your body absorb the calcium that is already in the milk. By fortifying milk with more vitamin D than it has naturally, the milk company gives you a rich source of vitamin D and makes the calcium in the milk more easily absorbed by your body.

A **vitamin deficiency** occurs when your body doesn't get the proper amount of vitamins it needs. Though people can survive with some vitamin deficiencies, they are at risk of developing really big problems. For example, if someone has a vitamin A deficiency, it will first show up as night blindness, making it hard for that person to drive at night. Untreated, this vitamin deficiency might eventually result in complete blindness.

Thankfully, the Lord designed our bodies to store many vitamins. That way, they can be called into action when they are needed. Do you remember which organ stores a lot of vitamins? Your liver! Some vitamins, however, are not stored in the body. It all depends on whether the vitamin is **fat-soluble** (sol' yuh bul) or **water-soluble**. In other words, does it dissolve easily in water or in fat? Vitamins that dissolve in

Although bruising after an injury is normal, excessive bruising may be a sign of a vitamin deficiency, such as a lack of vitamin C.

water are not stored by your body. What's not needed is just removed. The fat-soluble vitamins, however, are stored in your body's fat.

Many vitamins are named after the letters of the alphabet, while some have other names. The "alphabet" vitamins are: vitamins A, B (there are eight different kinds of B vitamins), C, D, E, and K. Let's explore a few different vitamins, and then we will take a look at minerals.

Vitamin A

Vitamin A is found in foods like milk, eggs, cheese, liver, pumpkin, peaches, spinach, sweet potatoes, mangoes, and carrots. This important vitamin keeps all your smooth muscles running efficiently and fights any infections that try to take over your body. It also helps the nerves in the eyes so you can see better. People who don't get enough of it can have trouble fighting off infections caused by even minor germs. But, don't think that taking a lot of vitamin A will make you better off! You see, vitamin A is a fat-soluble vitamin, which means that your body can store it. If you regularly get way too much vitamin A over a long period of time, your body continues to store it, and it continues to build up in your body. If it builds up enough, it can cause hair loss, dry skin, weak bones, and headaches.

Vitamin C

Vitamin C is a water-soluble vitamin, which means that any excess you take in (beyond what your body needs at that time) will come out in your urine. So it's important to get regular amounts of vitamin C, which is found in most uncooked fruits and vegetables. Unfortunately, it is destroyed by high heat, which happens when food is cooked. Citrus fruits are an excellent source of vitamin C, as are green peppers, strawberries, tomatoes, and green vegetables. Because vitamin C is water-soluble, it is difficult to get too much of it. Your body doesn't store it, so it doesn't build up in your body.

Your body requires vitamin C for the growth and repair of your tissues. Remember collagen – a substance which is in parts of your bones, ligaments, and tendons? Well, without vitamin C, your body cannot make collagen. Vitamin C also helps burns and wounds heal, keeps your teeth and gums healthy, and it is believed to help speed the healing of colds and other illnesses.

This painting shows Dr. James Lind caring for sailors with scurvy. His experiments led him to conclude that fruits like limes could prevent the dreaded disease among sailors.

Where's the C at Sea?

Long ago, when men sailed the oceans for long periods of time, they often suffered from a disease called scurvy, which we now know is caused by a vitamin C deficiency. When someone goes for several weeks without any vitamin C, scurvy begins to set in. On their long journeys, the sailors had mostly dried meat, biscuits, water, and rum. Without vitamin C, they couldn't make collagen. Without collagen, cuts didn't heal, teeth loosened, gums bled, tendons decayed, blood vessels and muscles broke down, and eventually the person died.

A doctor on a British naval ship (James Lind) saw men suffering from this disease. Aware of their terrible diet, the doctor tried treating them with a number of different foods. He found that citrus fruits or citrus juices caused their symptoms to begin disappearing. From then on, sailors on long voyages were given some lime or lemon juice along with their meals. Have you ever wondered why British sailors are sometimes called "limeys?" Now you know why!

Cut a banana into slices. Break up a vitamin C tablet and sprinkle it over half of the cut slices. Check back in an hour. What happened? When fruits are exposed to air, oxygen begins to break down the cell walls, and the fruits begin to turn brown. This is called oxidation (ok' sih day' shun). Vitamin C is an antioxidant. That means it helps to stop the oxidation process. This is why people add lemon juice (which contains a lot of vitamin C) to avocados and salads; it keeps them from turning brown.

A freshly-cut apple is not brown like one that has been cut open for a while because it has not yet experienced much oxidation.

Vitamins D and K

Food is not your only source of the fat-soluble vitamins D and K. Your body can actually make these vitamins with a little help! You might remember that your body makes vitamin D to build bones and make them strong. Your body manufactures vitamin D when you are exposed to sunlight. You can also get vitamin D from dairy products, eggs, and fish. Do you remember that milk companies often add vitamin D to their milk?

People who don't get out in the sun enough can become vitamin D deficient. If you live in a warm climate, you need a couple of hours a week in the sun to make the vitamin D your body needs. If you live far from the equator, in a cool climate, however, you'll need a lot more because you usually cover most of your body when you go outdoors due to rain or cold. People who have a vitamin D deficiency often feel weak, experience muscle spasms, and have aches and pains in their bones and muscles. If they don't begin to get enough vitamin D, either from food, supplements, or sunlight, they'll get a disease called **rickets**. This disease affects how your bones remodel and grow.

Vitamin K allows your body to make chemicals that help your blood to clot, which keeps you from losing too much blood when you cut yourself. You'll learn more about blood clots in a later lesson. Vitamin K is also needed to keep bones and other tissue healthy. The first sign of a vitamin K deficiency is easy bruising. Do you remember all those little bacteria you have in your intestines? Well, they produce vitamin K for you. However, they really don't produce enough, so you'll have to get the rest by eating your spinach, salad, whole grains, potatoes, and cabbage.

B Vitamins

There are a number of B vitamins, and like vitamin C, they are all water-soluble, so you need to consume them every day. They are active in nearly every area of the body, and they especially help the cells make energy. B vitamins also help to keep your nervous system, blood cells, and skin working normally. Whole grains are rich in B vitamins. Refined breads and cereals have B vitamins added. B vitamins are also found in nuts, meat, broccoli, and cabbage.

Vitamins: The Inside Story

You've learned a lot about vitamins and what happens if you don't get enough of them, but what do vitamins actually do? Well, after you've eaten something rich in vitamins, the vitamins enter your blood and are taken to your cells. Within the cells, the vitamins act as helpers. They generally assist in tasks that would either go too slowly or not go at all without the help of the vitamin. For example, your body has all sorts of enzymes, which are special proteins that speed up chemical reactions that must happen quickly for your cells to function properly. Many enzymes need a helper in order to function. We call this helper a **coenzyme**. Some vitamins act as coenzymes, and other vitamins are used by your body to build coenzymes. So vitamins play a "helper" role in your body. Just like your chores get done faster (and usually better) when you have a helper, the chemical processes in your body work better and faster if they have their vitamin-based helpers.

Try to tell someone all you remember about vitamins before moving on to minerals.

Minerals

Our family has a game called "Twenty Questions." One person thinks of an item, which the others try to guess. The first question asked is, "Is it animal, vegetable, or **mineral** (min' ur uhl)?" What are minerals? What do you think of when you hear that word? Well, minerals are not living, and they are not produced by living things (like vitamins are). Minerals are usually found in the earth, rivers, lakes, streams, and oceans. Your body must have them to survive.

The green stuff on this rock is malachite, a copper-containing mineral. You would never eat malachite, but your body does need small amounts of copper.

We won't go over each one, but the minerals you need most are: **calcium** (kal' see uhm), phosphate (fos' fayt), magnesium (mag nee' zee uhm), **sodium** (so' dee uhm), potassium (puh tas' ee uhm), zinc, **iron**, copper, **iodine** (eye' uh dyne'), selenium (sih lee' nee uhm), and chromium (kroh' me uhm). You might recognize sodium. You can find it in sodium chloride, which is better known as table salt. Like vitamins, you need only small amounts of minerals in your diet. But if those small amounts are not there, what happens to your body isn't pretty.

Calcium, as you may remember, is found in dairy products, cabbage, salmon, spinach, and sardines. If you were to have a calcium deficiency growing up, your bones wouldn't grow or remodel correctly, and you would not have good muscle contractions or muscle tone. Adults that do not get enough calcium lose their ability to think, they become very weak, and their hearts don't beat properly.

An iron deficiency causes your blood to be less efficient at carrying oxygen to your tissues. This can result in troubled breathing, extreme exhaustion, and worse – risk of death. You'll understand why after you read the chapter on blood. Fortunately, iron is plentiful in beef, clams, turkey, chicken, seeds, beans, and leafy green vegetables.

Iodine is found in seawater. Since most of us don't live near the sea, we usually get iodine from eating table salt that has been fortified with this mineral. People who do not get enough iodine have problems with a gland in the neck called the **thyroid** (thy' royd). The problem results in a swelling of the neck called a **goiter** (goy' tur). Thyroid problems can also lead to a lower intelligence.

Minerals keep our blood, bones, teeth, and muscles working properly. If you eat a varied diet with whole grains, vegetables, nuts, peas, beans, meats, and dairy products, you'll get all the minerals you need. So do you see now why it's important to eat a varied diet? Eating a "varied diet" means you should eat lots of different kinds of things. Often families get stuck in a rut and eat the same things over and over again. However, a varied diet of healthy foods ensures that you can get the vitamins and minerals you need. Whole foods are the best for your body. Do you remember the difference between whole-wheat bread and white bread? Whole-wheat bread uses all of the wheat grain, while white bread uses only a part of it. The more whole foods you eat, the more likely you are to get all the vitamins and minerals God designed the world to provide you.

It's often easier and more convenient to take vitamin and mineral supplements instead of eating a healthy diet. But many scientists believe that the vitamins found in foods are more effectively used by your body than the vitamins you may take in pills. Some scientists think this is so because there are many, many substances in foods that have not yet been identified that enable your body to better absorb and use the vitamins in the food you consume.

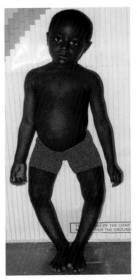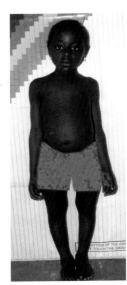

This child was suffering from a condition called rickets because of a lack of calcium in the child's diet. This caused the child's legs to bow out because the child's bones were not strong enough to hold the body's weight. After two years of being given calcium, the child's bones became stronger and the child recovered from rickets. (photos courtesy of Dr. Tom D. Thacher, M.D.)

Many families encourage their children to take vitamin and mineral supplements as an "insurance policy" to make sure they are getting enough of what they need. However, remember that taking supplements can't cover for an unhealthy diet. If you eat a lot of sugary foods, trans fats, and low-fiber foods, supplements won't "fix" the problems that come with an unhealthy diet. Also, if you take vitamin supplements, you should follow the directions on the bottle carefully. If you take too much of a vitamin, that can lead to problems, too.

What Do You Remember?

How do you know if you are dehydrated? Why do simple carbohydrates give you quick bursts of energy, while complex carbohydrates don't necessarily do that? What do carbohydrates change into inside the body? Proteins are made of what kind of molecules strung together? What is a complete protein? Which foods provide your body with omega 3 fatty acids? Name three vitamins that are important to get, and tell why they are important. Where are minerals found? Name two minerals, and tell why they are important for your body.

Notebooking Activity

Today you are going to create a one-week dinner menu that includes all the foods you need for a very healthy diet. You will also include nutritional information at the bottom of each daily menu, explaining what nutrients are contained in each food and how they help your body. Begin by studying the different nutrients you need each week. Then, look through cookbooks or recipe websites and decide which foods you will put on your menu for the week. After you have created your menu, you can go shopping with your parents. Ask if they'll let you help them prepare the special dinners you have designed.

If you have the *Anatomy Notebooking Journal*, you will find pages for these activities.

Project

For your project this week, you will create a food illustration based on the government's recommended daily allowance (RDA). Go to the course website I told you about in the introduction to the book. Find the link to the USDA's food recommendation and go there. This recommendation is based on the latest scientific understanding of nutrition.

Draw the illustration for your notebook and color the different sections. Be sure to include a list of the foods recommended for each colored section. Next, draw seven blank food illustrations. You will use these to chart the foods you eat over the next week. Using a new illustration each day, chart everything you ate that day. At the end of the week, compare the foods in your illustrations to what is recommended. Are you making healthy food choices for your body?

Experiment
Testing for Vitamin C

Which fruits contain the most vitamin C? You can test that for yourself.

Getting all the vitamins you need from the food you eat can be very tricky. We are told that oranges are full of vitamin C, and you probably know that drinking orange juice will give you vitamin C. But are there other fruits that have more vitamin C than oranges? Do different types of orange juice have more vitamin C than others? These are things you can learn through experimentation. Today, you are going to do an experiment that determines which fruits have the most vitamin C. You can also redesign this experiment to test different brands of fruit juice against freshly-squeezed juice. Whether you do this experiment as it is written or by testing other items, it is very important that you make your measurements carefully. This is one of those special experiments that won't work if you do not use the correct materials or measure them out exactly.

You will need:

- A Scientific Speculation Sheet
- Juice, freshly squeezed from different fresh fruits or vegetables that you think might contain vitamin C (oranges, tomatoes, strawberries, peaches, etc.)
- Cornstarch
- A 2% iodine solution (available at most drug stores)
- A medicine dropper
- A juice glass
- Several small cups or test tubes
- A measuring cup
- Measuring spoons
- A small pot
- A stove
- A spoon

In order to do this experiment, you need to make a solution that can test for vitamin C. Since this kind of solution indicates the presence or absence of vitamin C, it is called a "vitamin C indicator." You will add drops of fruit juice to this solution to test for vitamin C. The indicator will change color to show how much vitamin C is present in a fruit. If the color lightens a great deal, or fades away entirely, that means a lot of vitamin C is present in the fruit. If it changes only a little, only a little vitamin C is present. If it does not change at all, no vitamin C is present.

To make the indicator:

1. In a small cooking pot, mix 1 cup of water with 1 tablespoon of cornstarch.
2. Bring the solution to a slow boil and continue cooking for 5 minutes, stirring while it cooks.
3. After 5 minutes, remove the pot from the stove and let it cool.
4. After the cornstarch solution has cooled, measure out exactly ⅓ cup of water and put it in the juice glass.
5. Using a medicine dropper, add 10 drops of the cornstarch solution to the juice glass. You are now finished with the cornstarch solution, but it can be stored in the refrigerator for a few days if you want to make more indicator later.
6. Rinse out your medicine dropper.
7. Use the medicine dropper to add 15 drops of iodine to the juice glass.
8. Stir the contents of the juice glass until it changes to a dark blue/violet color. If the indicator is not a very dark blue, add another drop or two of iodine.

Your indicator is now ready to be used. You have made enough vitamin C indicator to run four tests. For each test you will need exactly 4 teaspoons of indicator. Measure the indicator into four small cups or test tubes and label each with the name of the fruit you will be testing. If you want to test more items, just make more batches of indicator.

To test for vitamin C:

1. Wash the fruit and cut it open.
2. Squeeze two tablespoons of juice from the fruit into a small, clean cup or bowl.
3. Using a medicine dropper, add exactly 10 drops of juice into the cup that is labeled for the fruit and contains the indicator solution.
4. Swish the indicator around and watch carefully for a color change.
5. Put the color change you see into one of these four categories: dark blue/purple (which means there was no change), light blue, pink, light pink, or clear. Record the category you observed.

6. Carefully wash the medicine dropper and repeat for each new fruit that you want to test.
7. When you have tested all the fruits, arrange the solutions in order from lightest to darkest. The lighter the indicator, the more vitamin C is present in the fruit.
8. Record the results on the Scientific Speculation Sheet. Which fruits contain the most vitamin C? Were there any that contained no vitamin C at all?

You can repeat this experiment and test just one fruit, testing for the differences in the amount of vitamin C in freshly-squeezed juice, stored juice, or processed juice. For example, you can test freshly-squeezed orange juice against frozen orange juice, bottled orange juice, or an orange sports drink. You could also test whether orange juice loses vitamin C over a period of days. There are many variations. Be creative! By the way, this is a great science fair project!

Fruits and vegetables are important components of a healthy diet.

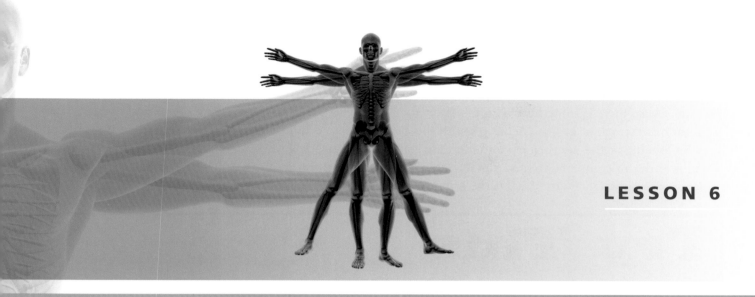

the
RESPIRATORY SYSTEM

How many breaths do you think you take in one day? Let's find out. Get a timer and try to breathe as normally as you can for one minute. Count every time you take a breath. Most kids breathe about 18 times a minute. How many times did you breathe in one minute? To find out how many times you breathe in one day, multiply the number of times you breathe in one minute by 1440, because that's how many minutes there are in a day.

What if you had to think, concentrate, and try to breathe every single time you needed a breath? Well, you probably couldn't do anything productive with your day. You would be so busy breathing, it would take all your time! Yet, God designed your **respiratory** (res' pur uh tor' ee) **system** to include your nose, trachea, and lungs so that you could breathe without thinking about it. Of course, you can also control your breathing somewhat. In other words, you can voluntarily hold your breath, at least for a little while. You can also pause and take a deep breath if you think about it. This is especially good when you are angry, anxious, about to go under water, or when you

Even though you don't have to think about breathing, you can control it, like when you take a deep breath before going under water.

103

experience an unexpected scare! Taking that extra breath gives you the extra oxygen you need.

Oxygen isn't the only thing in the air you breathe. In fact, it's only about 21% of what is in the air. Air is made up of mostly nitrogen (about 78%) and a tiny bit of many, many other gasses, including carbon dioxide. Also, there is sometimes a lot of water vapor in the air. We call it **humidity** (hyoo mid' ih tee). The amount of water vapor in the air changes with the weather, so you can usually find out how humid it is in your area by looking at the weather report. Humidity is usually listed as a percent that tells you how much water vapor is in the air compared to how much could be in the air. If the humidity is 100%, the air has as much water vapor as it can possibly hold. What is the humidity right now where you live?

Try This!

Go to a website that posts the daily weather (try http://www.weather.com or check the course website we told you about in the introduction). See if you can discover the humidity where you live. Now look at a place that typically has very little humidity, like Phoenix, Arizona. What is the humidity there today? Now find a place that has a lot of humidity, like Key West, Florida. Do you see a big difference in the humidity levels? It depends on the specific weather that is going on today, but you will usually see a difference.

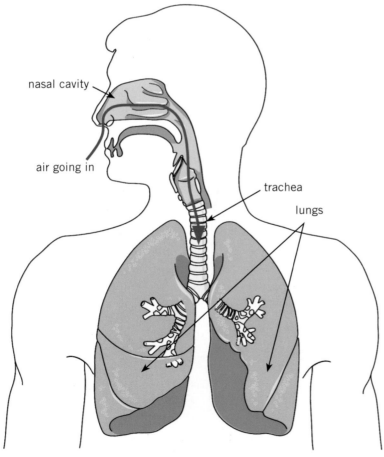

nasal cavity

air going in

trachea

lungs

Most of the air that gets to your lungs passes through your nose.

There are other things in the air besides water that are not so great for our environment – or for our bodies. They include dust, viruses, bacteria, pollen grains, fungi, and many different chemicals. When you inhale, you're inhaling some of these potentially harmful things along with the oxygen you need to survive. So what happens when you breathe in all this stuff? Well, as you probably already know, it all goes toward your lungs. But then what? Let's find out!

Most of the time when you breathe, air is taken in through your nose. If you've ever had a cold and had to breathe through your mouth for a long time, you probably remember that it wasn't very comfortable. Why don't you try that now: take four deep breaths through your mouth. Now, take four deep breaths through your nose. Which felt more comfortable? God designed you so that most of the air you take in passes through your nose. That's because your nose leads to your nasal (nay' zul) cavity, which has been specially designed to filter out a lot of the dust, pollen, bacteria, and other stuff in the air. Your nose also works to warm and moisturize the air you breathe. This makes the air safer to breathe and more comfortable as it enters your lungs. You see, God placed very delicate cells in your lungs that need to be protected. Your nasal passages and the tubes that go to your lungs help to protect them. Let's take a look at how this works.

Hairy Catchers

Did you know that you have hair in your nose? Just in case you don't believe me – it's time to get a mirror! Use the mirror to look up your nose and see if you can identify the tiny hairs inside. This may seem like a funny place for you to grow hair, but God put that hair in your nose for a good reason. Nose hairs trap the large particles of dust that you breathe in every single day, and that protects your lungs.

Mucky Mucus

In addition to hair, God designed some other safeguards to keep dust and other bad things out of your lungs. One of these safeguards is called **mucus** (myoo' kus). Mucus is extremely important because it's sticky, so the stuff that doesn't belong in your lungs sticks to it. Did you know that God made you with special kinds of tissue that produce mucus? They're called your **mucous membranes**, and they are found in many parts of your body, including your nose and all the tubes that lead to your lungs. Let's try a little experiment to see how mucus works. This will take a few days.

> ## Try This!
> Get some honey and two pieces of cardboard. Put a coat of honey on one of the pieces of cardboard and put some water on the other. Now put both pieces of cardboard outside, preferably on a table or other surface to which insects won't have easy access. Check the cardboard pieces each day for three days. Remember, mucus is sticky, like honey. You will soon see from this experiment how important mucus is for your body.

Slashing Cilia

In addition to mucus, God placed tiny motors in your nose. It's true! Isn't that amazing? There are teeny tiny motorized whips waving back and forth on the cells lining the back of your nose. These whips are called **cilia** (sil' ee uh), and they are constantly moving the mucus that has particles stuck to it towards your mouth where it can be swallowed and then destroyed by your stomach acid. Have you ever wondered why your nasal mucus goes down your throat? It's the cilia in action!

Believe it or not, the air you breathe has more done to it than just being filtered by your nasal hair and mucus. In addition, it needs to be conditioned with moisture and heat. Indeed, God designed a perfect air-conditioning system right inside your head! He did this because the air you breathe is usually too dry and not at the right temperature. If you've ever run outside on a cold day, you may remember how taking deep breaths through your mouth made your lungs burn. This is because inhaling cold, dry, or contaminated air damages the sensitive cells lining your lower airway. You see, your body is warm inside, almost 100 degrees.

These children are skiing in dry, cold weather which would damage the cells lining their lower airways if their noses did not work properly. God designed your nose to moisten and heat the air you breathe so that it is just the right temperature for your lungs. The remarkable design of your nasal cavity, with its own air conditioning system, points to an amazing God who thought of everything when creating you.

So unless the air is close to 100 degrees, it's got to be warmed up before it's sent into your lungs. Also, because your lungs are really moist, dry air hurts them. So in addition to being warmed, the air also needs to be moisturized. It's wonderful how God has designed your body to do all of this.

Crazy Conchae

As the air travels up your nose, it hits your **conchae** (kong' kee). Have you ever seen a conch shell? Well, that's what your conchae were named after. That's because a conch shell passage has twists and turns, just like the passages formed by the conchae in your nasal cavity.

God had a special reason for creating your nasal cavity this way. Your conchae interrupt the air flow, making it travel like a twisting roller coaster – going this way and that way, slamming against the mucus in the nasal cavity to make sure dust is removed. Sounds fun, doesn't it? As the air hits the walls of your nose, it also gets heated by the warm tissue found there!

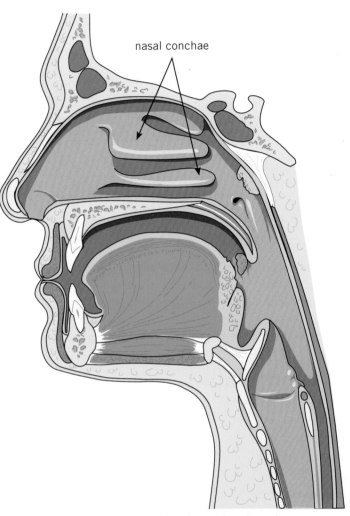

nasal conchae

Your nasal conchae make sure the air passing through your nasal cavity is cleaned, moistened, and brought to the correct temperature.

Place your hand about eight inches from your face. Now blow on your hand. The air is a bit cool, isn't it? Keep your hand there and hold the other hand over your mouth, forming a small circle. Blow through the circle to the hand that is in front of you. You may have to adjust the size of the circle to get the air to hit your hand. What happened to the temperature of the air?

The heat from your hand warmed the air in the same way that the air hitting the tissue in your nasal cavity gets warmed by your body. This happens even more effectively in your nose, because the lining of your nose has many more blood vessels than your hand. Have you ever gotten a bloody nose? Your nose bleeds easily because there are many tiny blood vessels in the mucous membranes. The blood in those vessels is at body temperature. All that blood flowing near the lining of your nasal cavity helps to warm the air you breathe if it is too cold or cool it if it is too hot.

Holes in Your Head

Have you ever heard the phrase, "You need that like you need a hole in your head"? The person saying this is trying to make the point that you don't need whatever it is that's being offered. However, it's not such a good phrase, since *you actually do need holes in your head.* Without them, you would have a lot of trouble.

You see, after air passes through the nasal cavity, some of it travels into the holes in your head called your sinus cavities. These holes help make your head lighter, and they continue the warming, moisturizing, and filtering that began in your nostrils. A lot has to take place to get the air ready for your lungs.

Like all mucous membranes, the membranes lining your sinus cavities can become infected. When unhealthy bacteria or viruses infect the sinus walls, your body begins to produce enormous amounts of mucus. We call it a runny nose. Have you ever had a runny nose? Most likely, it was caused by something up in your sinuses that your body didn't like. The mucus was trying to trap and send away the unwanted intruders.

If the intruder is not removed by the runny nose, it can cause problems in the mucous membranes, and an infection can occur. Infections can be painful and often make you feel sick. If you've ever had a sinus infection, this is what happened. Sinus infections sometimes do not go away on their own, especially if you're not eating nutritious food or are under a lot of stress. In this case, they must be healed with antibiotics. If left untreated, a sinus infection can spread, so don't ignore the symptoms if you have an infection.

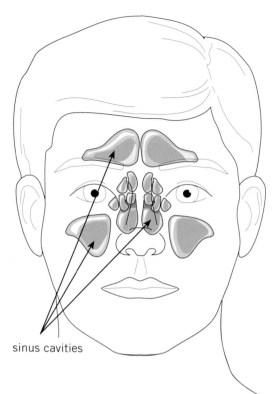

sinus cavities

Your sinus cavities are actually holes in your skull.

Have you ever wondered why you sneeze? Sneezing happens when your nasal passage gets irritated. Sometimes the irritation is caused by a virus that has infected your tissues, like what happens when you catch a cold. A sneeze might also happen when a large particle gets trapped in the mucus that lines your nasal cavity. Well, God created an amazing thing that happens when something is in your nose that doesn't belong there – suddenly a rush of air, flying at about 100 miles per hour, comes crashing through your nasal passage, traveling out your nose and mouth. That's a hurricane-speed wind, and your body can produce it! This strong wind can send millions of tiny particles (including viruses) into the atmosphere, where they can spread around.

So, let's get back to breathing. I know you've been doing it this whole time, but let's study it a bit more. So, you've already taken the air into your hairy nose; it has gone through the conchae and up into your sinus cavities. From there, it travels down, down – far down your neck through your **pharynx** (far' ingks).

The pharynx has three parts. The uppermost part is the **nasopharynx** (nay' zoh far' ingks). Have you ever ridden in an airplane and felt your ears pop? You may have felt this when diving to the bottom of a deep pool or lake. This actually happens in your nasopharynx. It occurs because the middle ear is connected by a

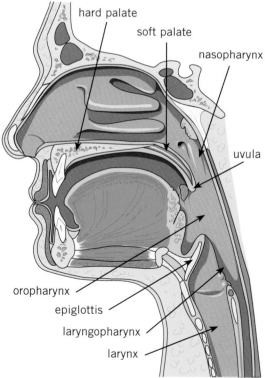

hard palate
soft palate
nasopharynx
uvula
oropharynx
epiglottis
laryngopharynx
larynx

Air passes through the three parts of the pharynx down to the trachea.

107

tiny tube to the nasopharynx, and when your ears need to equalize the pressure between the middle ear and the atmosphere, your ears "pop."

The nasopharynx ends at the soft palate, which we discussed in the lesson on the digestive system. The soft palate rises to close off the nasal cavity when you swallow food or liquid, so that these substances don't travel upwards into your nose. The uvula is visible if you open your mouth wide and look in the mirror. It is a small pink projection hanging downward from your soft palate. The uvula marks the lower edge of the nasopharynx.

Try This!

With a mirror in hand, open your mouth wide to view your uvula. You'll have to flatten your tongue a bit. Now try to swallow and see what happens to your uvula. It should flatten out and help close off your nasal passage. If you can't see this happening, it may be because your tongue is getting in the way as you swallow. Try it again.

Just below the nasopharynx is the **oropharynx** (or' oh far' ingks). "Oro" is like the word "oral," which refers to the mouth. So, it's pretty easy to remember that the oropharynx is the part of the throat that is closest to the mouth. When you swallow food, it passes from your mouth to your oropharynx and then down to your esophagus. If you remember the epiglottis from the lesson on digestion, you'll know that this special flap of tissue closes off the airway when you swallow, directing food away from the **larynx** (lar' ingks) and into the esophagus. The oropharynx ends at the epiglottis.

The last part of the pharynx is the **laryngopharynx** (luh ring' go far' ingks). Whatever is in the pharynx must travel either down the esophagus and to the stomach, or down the larynx and to the lungs. Although "larynx" may sound a bit like a Doctor Seuss character, it's actually the part of your body that gives you the ability to speak. Because of this, some people call your larynx your voice box. You can feel your voice box by placing your fingers just behind your Adam's apple (the little bump in your throat).

Speaking Strings

If you pull a string taut and then pluck it, you will hear a sound. If you put that string over a hole, the sound will actually echo. This is similar to how God designed your voice box. Your larynx has a pair of "strings" that can vibrate as air blows by. The "strings" in your larynx are called **vocal cords**, and they are strips of tissue that you can pull tightly across your larynx in order to make sound. Close your mouth and make a humming sound. You are sending air up from your lungs, through your larynx, and across your vocal cords, which you pulled tight across your larynx. This causes them to vibrate. The more air you send up, the more your vocal cords vibrate, and the louder it sounds, as long as your vocal cords are stretched across your larynx. When you relax your vocal cords, they spread apart out of the airway so that they do not vibrate and you can breathe silently.

To make high-pitched sounds, you tense up your vocal cords, making them tighter. To make low sounds, you relax them a bit. Why do you think men have lower voices than women? It has to do with the size of their vocal cords. Men tend to have longer, thicker vocal cords, which produce lower voices. The more high-pitched your voice, the shorter and thinner your vocal cords. As boys grow up, their vocal cords get longer and wider, making their voices lower.

Try This!

Stretch some thin and thick rubber bands over the opening of a plastic container. Pluck the rubber bands to hear the different pitches they make. Do the thicker ones make lower or higher pitches? What happens if you stretch the rubber bands even tighter across the mouth of the container? What happens if you loosen them?

No matter what size vocal cords you have, no one could hear you without some sort of amplifier built into your body. When we want to amplify (increase the volume of) the sounds we hear, a cavity (a space) is usually needed unless we are using electrical devices. This is how an acoustic (non-electric) guitar amplifies its sound. Have you ever shouted into a cave or a huge empty room? The empty space causes echoes and makes your voice sound louder. Well, this is another thing that the empty spaces in your head do for your voice. The empty spaces in your chest, mouth, and skull amplify your voice so that people can hear you when you speak. People with larger chests, mouths, and nasal passages typically have louder voices. Of course, you can also increase the volume of your voice by just forcing more air through your larynx. The more air passing over your vocal cords, the louder the sound they produce.

Did you know that people who smoke develop lower voices? This is because smoking causes the vocal cords to become inflamed and swollen. Also, smokers are unable to speak with much volume because their lungs shrink significantly, causing less air to flow from them. I'm sure you've heard that smoking cigarettes can increase your risk of lung cancer. Did you know that smoking can also increase the risk of cancer in the larynx? If this happens, the larynx must often be removed to prevent the cancer from spreading.

Try This!

While your vocal cords produce the initial sounds you make, your lips and tongue form those sounds into words. Try to talk while holding your tongue. Now try to talk without moving your lips. This time, try to talk without moving your jaw. If you promise to be very careful not to swallow it, try to talk while holding a grape under your tongue. Do you see how important your lips and tongue are for actually forming the words you want to say?

What have you learned so far? Tell someone so that you can remember it well.

Trachea Track

After the air passes through your larynx, it enters your trachea. Feel the front of your neck again, this time below the voice box. Do you feel those ridges? They are actually cartilage rings that are wrapped around your trachea. They may feel a little strange, but without these rings around your trachea, you wouldn't be able to survive. You see, God created them to keep your trachea from collapsing. Let's see how this works.

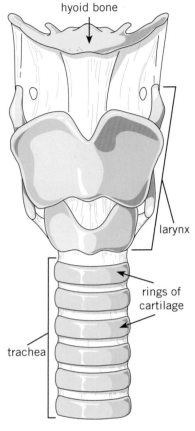

hyoid bone

larynx

rings of cartilage

trachea

Try This!

Get a straw and blow air in and out of it, as if it were your trachea. Now, pinch the end of the straw with your fingers to keep it closed. What happens to the straw when you try to suck the remaining air out of it? The pressure difference between the air outside of the straw and the air inside the straw causes the straw to collapse. This is exactly what would happen to your trachea if God had not placed cartilage rings around the entire length of it. This is because inspiration (breathing in) occurs when the pressure in your chest falls. This drop in pressure pulls in air from outside your body, the same way sucking on a straw will pull milk from a glass into your mouth. We'll get to what makes the pressure in your chest drop in just a minute.

Bronchi Branches

Like an upside down tree, the trachea branches out into two tubes called **bronchi** (brong' kye). Have you ever heard of **bronchitis** (bron kye' tus)? Well, the word part "itis" often refers to something that is swollen or inflamed. Bronchitis occurs when your bronchi are swollen, usually because of an infection. People with bronchitis cough a great deal because of the irritation it causes and to remove excess mucus.

The trachea divides into two bronchi, one branching out to the right and the other branching out to the left. Each takes air to a lung. Study the picture and see if you can find any differences between the right lung and the left lung.

You might have noticed that the right lung has three lobes, while the left lung has two lobes. Why? Well, remember that the heart is in the same area as the lungs (see the drawing on page 38). In the human chest, the heart lies more on the left side than the right, so there is less room for lung tissue on the left side. That is why there is one less lobe in the left lung.

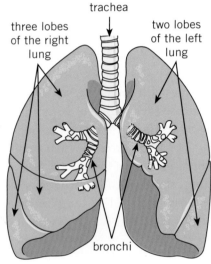

trachea

three lobes of the right lung

two lobes of the left lung

bronchi

In anatomy, "right" and "left" refer to the person to whom the organ belongs. So the left lung in this drawing is on your right.

Baby Bronchioles

So the air goes down your trachea and into your bronchi, but then your bronchi continue to divide over and over again so that there are many, many tubes running throughout your lungs. Every time the bronchi divide into more bronchi, the tubes get smaller. Eventually, they get so small that they are called **bronchioles** (brong' kee ohlz), which is a Latin way of saying "little bronchi." These are very small, thin-walled tubes that carry air to where the lungs can finally use it. There is another coughing illness called **bronchiolitis** (brong' kee uh lye' tis). It's similar to bronchitis, but in this case, the tiny bronchioles are inflamed, not the bronchi.

The trachea, bronchi, and bronchioles are filled with air and lined with mucus and cilia. The mucus usually catches anything that gets past the cleaning processes in your nasal cavity. The cilia push the mucus (and anything stuck to the mucus) upward toward your throat, where it can be swallowed. If your body is making a lot of mucus, your cough reflex will kick in to help expel the mucus from your lungs.

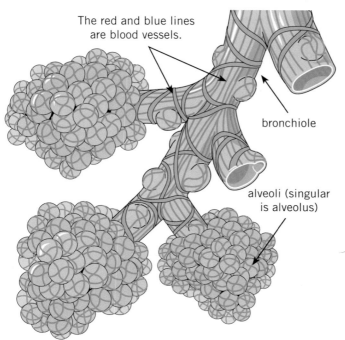

The red and blue lines are blood vessels.

bronchiole

alveoli (singular is alveolus)

The alveoli are where the real action in the lungs takes place.

Start

Alveoli Alley

There are about 30,000 bronchioles in each lung. The bronchioles fan out to create little spaces called **alveolar** (al vee' uh lur) ducts, where the air enters clusters of little balloon-like sacs called **alveoli** (al vee' uh lye). Your two lungs hold about 600 million alveoli.

Remember how we said the trachea branches into something that looks like an upside-down tree? Well, the alveoli are like little berries at the very end of each branch. The alveoli may be tiny, but they are the most important part of the respiratory system. They actually get the oxygen you need out of the air and into your bloodstream. They also get the waste (carbon dioxide) out of the bloodstream and into the lungs so it can be exhaled. We'll discuss how it works in a moment. Before we do that, however, let's take a look at some problems we may encounter in this portion of our lungs.

Catching Cold

Have you ever wondered what happens when you catch a cold? Well, colds are caused by viruses that spread quite easily. Every time you cough or sneeze when you have a cold virus, you send out millions of droplets containing viruses that can infect anyone who touches the droplets before they dry out. This is how it works: someone touches the droplets then touches his eyes, nose, or mouth. The viruses then enter the cells, especially those lining the mouth and nose, and take over. It's a hostile takeover, too! Once inside the cells, they use the cells' own protein-making systems to make more viruses. This would be like a foreign army coming into your country and making all the citizens work for them in order to become an even stronger army. Now you know why it's important to sneeze into a tissue and wash your hands when you have a cold!

When this virus invasion happens, your body sends special white blood cells to fight the infection. Blood vessels in the area swell when the fighting starts. This swelling causes fluids to be secreted – remember the runny nose? The swelling can sometimes cause pain, especially if it's your throat that is swelling. If all goes well, your body's disease-fighting system will conquer the virus. Amazingly, there are hundreds of different viruses that can cause colds. Given the fact that we're out and about each day, it's easy to see how people can continually catch new colds. There are also lots of other things roaming about that can cause even worse illnesses. Both viruses and bacteria can infect any part of the respiratory system. Given the billions of bacteria and viruses that are suspended in the air, it's truly miraculous that we aren't sick a lot more often. We'll learn more about our immune system (the system that fights sickness) in another lesson.

Asthma Attack

Do you know anyone who has **asthma** (az' muh)? Asthma is a disease that makes it very hard to breathe at times, because the muscle around the bronchioles contracts and narrows them. In addition, inflammation can cause the tissue inside the bronchioles to swell, narrowing them further. This reduces the space through which the air travels. It's like the difference between breathing through a paper towel tube and breathing through a straw. It's a lot harder to get oxygen in and out at the rate we need when the space through which the air moves is small. Why do asthma patients' bronchioles get smaller? Well, it's usually a reaction to an allergen

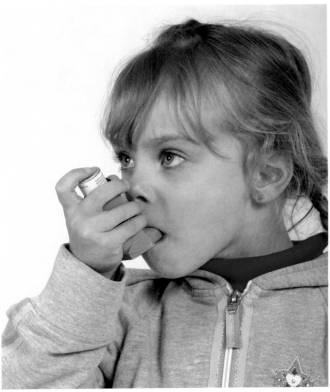

This girl is using her inhaler to help her breathe during an asthma attack.

that gets into the lungs. When the allergen arrives, the body begins a battle to fight it. Then, everything begins to swell – including the tissue inside the airways. Even a small amount of swelling can cause a dramatic reduction in oxygen and air supply. This can be a very dangerous situation to be in! Someone who is known to have asthma problems often carries an inhaler. When it gets difficult for the person to breathe, he or she inhales medicine from the inhaler. Typically, the medicine causes the bronchioles to get larger, allowing for more air to pass through them. Stop

Smoking Insanity

Why do people smoke? There are so many dangerous chemicals in one single cigarette and each one does something harmful to your body! As you can see from the picture below, smoking causes damage to the lungs. In addition, the carbon monoxide inhaled by smokers is poisonous to the body. It reduces the blood's ability to carry oxygen. This makes exercise, which requires oxygen, more difficult. Cigarettes have a chemical in them called **nicotine** (nik' uh teen'). The nicotine in smoke is addictive. That means as you use it, your body

begins to need it. That may seem odd, but it is yet another terrible side effect of smoking. Studies have shown that some people become addicted to smoking after a few cigarettes! Since people don't know in advance if they will become quickly addicted to nicotine, it's better not to take the risk of smoking even one cigarette. Nicotine also causes a person's blood pressure to increase. We'll learn about blood pressure in the heart lesson.

Remember all those special cilia in your mucous membranes and down in your lungs? They filter the air and work to keep you healthy. Well, smoking eventually destroys many of the cilia in your respiratory system, and they cannot beat out mucus and other

These two pictures compare a non-smoker's lung (left) to a smoker's lung (right). Look how dark and damaged the smoker's lung is!

irritants. This makes smokers much more likely to have serious and long-lasting infections. The damage from smoking can eventually make the alveoli break down and disintegrate. This is not as good as having many tiny alveoli. As a result, the lungs are less efficient, so the smoker must breathe in and out more often just to get the oxygen he needs.

Explain all that you have learned in your own words before moving on.

The Great Exchange

Are you still breathing? Good. Let's return to the discussion about what happens to the air once it goes down into your lungs and into your alveoli. Each single alveolus is covered with tiny blood vessels whose walls are only a single cell thick. As blood passes through these tiny vessels, oxygen seeps from the alveolus through the wall of the blood vessel and into a blood cell. In addition, blood cells that are carrying carbon dioxide (remember, that's waste they picked up from the tissues) give up that carbon dioxide, and it travels through the wall of the vessel and into the alveolus so it can be exhaled. So in the alveolus, the blood "trades" carbon dioxide for oxygen. You see, the blood that comes to your lungs is low in oxygen, having given up its oxygen to the tissues that needed it. So, it comes back to your lungs for more oxygen, exchanging it for the waste it picked up along the way. Oxygen in, carbon dioxide out. Doesn't that make you want to take a deep breath?

Filled to Capacity

If an adult man were to breathe in as deeply as he could, he would fill his lungs with about 6 liters (that's just over 6 quarts) of air. This is called his **total lung capacity.** Some people have a greater total lung capacity than others, which means they can breathe in a larger volume of air. Men usually have a greater total lung capacity than women. However, people rarely breathe that deeply. Instead, most people have about 2-3 liters (just over 2-3 quarts) of air in their lungs at any given time, at least when they are not exercising.

Interestingly enough, you cannot exhale all that air. When you aren't exercising, you generally only exhale about one-half of a liter of air. This means that somewhere between 70% and 80% of the air you normally have in your lungs stays there while you exhale. If you think about it, that's a good thing. After all, blood runs through your alveoli all the time, whether you are inhaling or exhaling. So it is good that there is still a lot of air in your lungs when you exhale, because that means the blood running through your alveoli still gets oxygen, even while you exhale. So does that mean you have a lot of old, musty air in your lungs? Of course not! Even though you only breathe out a small fraction of the air that is in your lungs, the air that remains is mixed with the new air that you breathe in.

You can increase your total lung capacity through exercise. We'll do an activity at the end of this lesson that will increase your total lung capacity in only one month. Now it is important to realize that because we almost never breathe in as much as we possibly can, your total lung capacity really isn't all that important. However, if your total lung capacity goes up, the amount of air that is typically in your lungs goes up as well. As

The more you exercise, the longer you can hold your breath, which comes in handy when you are swimming!

a result, you have a greater reserve of air in case you need to hold your breath for a while. Thus, exercise does benefit your respiratory system. It allows you to increase the amount of air in your lungs, which will help you to hold your breath longer and exercise harder and longer as well.

Now that we know what happens to the air once it enters your lungs, let's explore the mechanics that get air into your lungs. But first, let's do a little exercise to learn something first hand.

Try This!

Take a roll of transparent tape (like Scotch tape) and pull out a long strip. Wrap the tape around your chest while you are exhaling, and secure it while you hold your breath after you have exhaled as much as possible. Now take a deep breath. What happens to the tape? It stretches, doesn't it? It will probably break as you inhale, because your chest gets larger when you inhale. Do you know why your chest gets larger? Where does the air come from? Remove the tape, and we'll learn all about it.

As you know from our studies so far, the air is coming from outside of you, traveling through your nose and/or mouth and down into your chest. All of this is accomplished by a long muscle below your lungs called your **diaphragm** (dye' uh fram').

Diaphragm Design

When your diaphragm is at rest, it curves upwards into the chest cavity. When you inhale, your diaphragm contracts, which flattens it out and pulls it down. Have you ever played with one of those large, colorful, round parachutes? Do you remember you and all your friends holding on to the parachute around the edges and pulling it tight? Well, your diaphragm is like the parachute. When it contracts and pulls tight, it flattens out. At the same time, muscles around your ribs contract, pulling the ribs both upwards and outwards. Both of these actions increase the volume of your chest cavity. This pressure gradient, with air pressure higher than pressure in the lungs, causes inspiration.

When you breathe out, your diaphragm relaxes and gets longer. The muscles of your ribs relax and allow your rib cage to fall inward and downward. This causes your chest cavity to decrease in size, forcing the air out. Breathing in (inhaling) is active, meaning you must use muscle and energy to do it. Breathing out (exhaling) is usually passive, meaning you usually don't have to work to do it – it is simply accomplished when the working muscles relax.

What about when you need to cough or when you want to blow out your birthday candles? At those times you'll find yourself tightening your abdominal muscles to further reduce the volume of your chest cavity, which forces even more air out of your lungs. This is

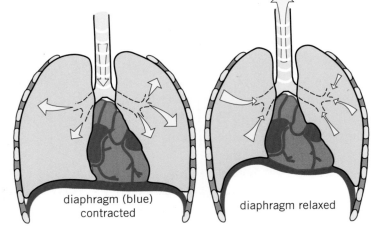

diaphragm (blue) contracted

diaphragm relaxed

When your diaphragm contracts, it flattens out, increasing the volume of your chest cavity. When it relaxes, it reduces the volume of your chest cavity, forcing air out.

called **forceful expiration**, and it not only uses the muscles in your abdomen, but also some of the muscles between your ribs. Some muscles between your ribs help the rib cage expand; others help the rib cage get smaller. Do you remember what muscles that work in opposite directions are called? Antagonistic. Once when I had a really bad cough, the muscles between some of my ribs contracted so hard that they tore a bit. For a week after, they hurt every time I took a breath. Luckily, that doesn't happen often.

Try This!

Put your hand over your belly, and blow out as hard as you can. Do you feel your abdominal muscles (the ones that tighten when you do a sit-up) contracting?

Heimlich Maneuver

Have you ever heard of the **Heimlich** (hym' lik) **maneuver**? Well, Dr. Henry Jay Heimlich, realizing that compressing the abdomen caused air to be expelled from the lungs, wondered if people could be saved from otherwise fatal choking events by manually compressing their abdomen to force the air (and hopefully whatever the person was choking on) out. Sometimes when a person laughs or is startled while eating, food sneaks into the trachea instead of going down the esophagus. Most of the time when this happens, the cough reflex kicks in, and the offending food is vigorously coughed out. But if the trachea is completely blocked, the victim cannot get air in to cough it out. This is where Dr. Heimlich comes in. He taught that compressing a person's abdomen would force air out of the lungs, which would expel blockages from the throat.

There are a lot of organs in the abdomen that can be damaged by doing the Heimlich maneuver improperly, so it's important to learn to do it correctly. You can learn to do it properly through the American Red Cross. It's a good thing to know how to do. Years ago my friend's two-year-old child was eating potato chips and walking around. She tripped and fell, choking on her mouthful of potato chips. As a result, she couldn't breathe, speak, or cough. The Heimlich maneuver sent that mass of chewed-up potato chips flying out of her mouth, and she was able to breathe again. After that experience, she didn't eat unless she was sitting down.

The Heimlich maneuver compresses the abdomen, forcing air out of the lungs.

Tasty Diaphragms

Have you ever eaten a **fajita** (fuh hee' tuh)? While you can get them with all sorts of different meats in the U.S., a fajita means something very specific in Mexico. It refers to a steak dish in which the meat is actually the diaphragm of a cow. A cow's diaphragm, like yours, is a thin sheet of skeletal muscle. It rests above the liver and beneath the lungs. Notice the placement of the diaphragm in the drawing on the right. Notice also how thin it is compared to most of the skeletal muscles you studied in Lesson 3.

So it's this thin sheet of muscle that is mostly responsible for your breathing. When it contracts, the pressure gradient between the outside air pressure and the pressure in your lungs causes inspiration. When it relaxes, it reduces the volume of the chest cavity, pushing air out of your lungs. This may sound complicated, but we'll do an experiment that shows you how this works. Now take a deep breath. Isn't it nice to know what's happening when you do that?

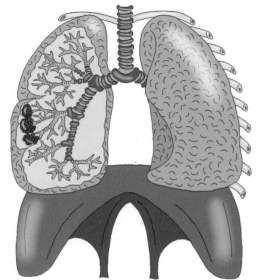

The diaphragm is a thin sheet of muscle under the lungs.

What Do You Remember?

What does the hair in your nose do? What does the mucus in your nasal passage do? What are cilia? Explain how the conchae help to warm and moisten the air you breathe. What are the thin strips of tissue in your larynx called? What determines your voice's pitch? What determines your voice's volume? How do the cartilage rings around your trachea help you? What are your bronchi? Explain the importance of alveoli. Name some of the dangers of smoking. How does the oxygen get from your lungs into your blood? What muscle is mostly responsible for your breathing?

Personal Person Project

Now you need to add a trachea, lungs, and a diaphragm to your Personal Person. If you have the *Anatomy Notebooking Journal*, there is a drawing for you to cut out. If not, use the drawing on page 115 as a guide to make your own.

Notebooking Activities

After writing down all that you've learned about the respiratory system, write a speech about the dangers of smoking. Give your speech to a group of friends or family members. After you do the experiment below, write down an explanation of the breathing process. Be sure to include drawings like the ones on page 114 so that your explanation is clear.

Experiment
Diaphragm Model

To help you to understand how the breathing process works, you will do an experiment that imitates it.

You will need:

- An empty plastic large-mouth drink bottle
- Scissors
- 2 balloons
- Tape

1. Cut off the base of the bottle near the bottom.
2. Place one balloon over the mouth of the bottle.
3. Push the balloon into the bottle so that the neck of the balloon remains over the mouth of the bottle.
4. Cut the bottom off the other balloon and use it to cover the bottom of the bottle. If you need to, secure the balloon to the bottle with tape.
5. Pinch the bottom balloon in the center and pull it gently away from the bottle. This creates a vacuum in the bottle and will cause the upper balloon to expand, filling it with air.
6. Release the bottom balloon. It should relax, forcing air out of the upper balloon.

   In this model, the balloon that hangs from the opening of the bottle represents your lungs, and the balloon covering the bottom of the bottle represents your diaphragm. When the diaphragm contracts, it creates a vacuum in the chest cavity, which causes the lungs to fill with air. When it relaxes, it pushes air out of the lungs.

Experiment
Vital Lung Capacity

You will need:

- An empty 2-liter plastic soda bottle
- A 1-foot-long piece of flexible tubing (like the kind you use for aquariums)
- A deep mixing bowl
- Tape
- Permanent marker
- Ruler

1. Place a strip of tape along the side of the bottle and use your ruler to mark off regular intervals.
2. Fill the bottle with water.
3. Fill the mixing bowl with water about 1/3 of the way.
4. Put your finger over the bottle opening and place it in the mixing bowl so that it is upside down (the opening is under water).
5. Add one end of the tube into the opening of the bottle.
6. Mark any air bubbles that may have entered into the bottle during steps 4 and 5.
7. While someone holds the bottle, inhale as you would normally inhale, put your lips around the tube and exhale as you would normally exhale. Don't try to exhale all of the air from your lungs; you'll do that in a moment. The water will spill out into the bowl, and the air from your lungs will be captured in the bottle.
8. Note the amount of air using your tape measuring system on the bottle. Write down that amount.
9. Refill the bottle and repeat the exercise, but this time inhale deeply. When you exhale, try to blow all the air out of your lungs. (You can't blow all the air out of your lungs.)
10. Measure where the air bubble is inside the bottle this time. This is the amount known as your vital lung capacity. Write down that amount.
11. You can repeat this experiment a few times and average your numbers. You can also wash off the tube and have your family members try the experiment too.

Save all of your materials because you will do a month-long project to increase your lung capacity, and then you will redo this experiment to see if you have succeeded.

Increasing Your Vital Lung Capacity

Now let's try to increase your vital lung capacity. Are you wondering how? Well, you can do this by exercising regularly. In this experiment you will jump rope, and you will start today. Start a timer and then start jumping rope at a reasonably fast pace. Go for as long as you can until you just can't do it anymore. See how long that was. Now…do this every other day for a total of four weeks, noting how long you can go each time. Try to keep the same pace every time you jump rope. Create a chart like the one at the bottom of the page to help you measure your progress.

　　You should probably see the amount of time you are able to jump rope increase as the days go on. The day after you finish your 4-week jumprope exercises, repeat the experiment and measure your vital lung capacity again. Did it change? It should have. The regular exercise should have increased it at least somewhat.

	Day 1	Day 3	Day 5	Day 7	Day 9	Day 11	Day 13	Day 15	Day 17	Day 19	Day 21	Day 23	Day 25	Day 27
Time														

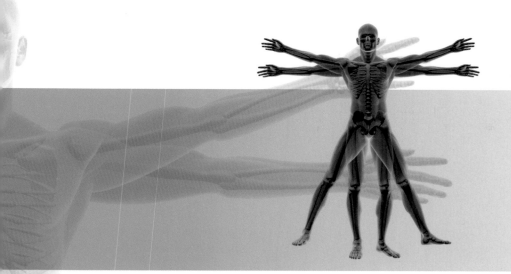

LIFE IN THE BLOOD

Although it may seem odd to spend a whole lesson discussing blood, believe it or not, your blood is one of the most fascinating things God created. It carries in it everything you need to live. It's simply astonishing how your blood flows through your body, reaching every cell to provide life, health, and healing. Of course, the Bible told us about this thousands of years before science discovered it. In fact, almost four thousand years ago, God spoke through Moses saying, "*For the life of the flesh is in the blood...*" (Leviticus 17:11).

This simply means that a person's life is found in the blood that courses through his or her body. Medical science has shown this to be true. When you are really sick, your doctor will often draw your blood to study it. By analyzing your blood, the doctor can often find out what medications you've taken, which of your organs are working properly and which are not, and what is causing your illness. Doctors can find out an enormous amount of information about your body's health just by studying your blood.

Do you realize that blood has eternal significance as well? It's true! Our eternal destiny depends upon blood. God's Word teaches us that the only way our sins can be forgiven is for someone or something to die

This is Joseph Aton Koch's interpretation of the animal sacrifice Noah's family made after leaving the ark. Animal sacrifice was necessary at that time for the forgiveness of sin.

and shed this life-giving blood in exchange for our sins. This is because sin must be punished, and unfortunately, that punishment is death (see Romans 6:23). You may be wondering, "What is sin?" Simply put, sin is "missing the mark" in our relationship with God. If we do something God says we shouldn't do, or if we don't do something God says we should do, we have sinned. In fact, the Bible says we have *all* sinned (Romans 3:23). Because God is merciful, however, He told His people long ago that an animal without blemish could die for their sins in their place. When Adam and Eve first sinned, an animal was sacrificed to cover their sin. Throughout the history of the Bible, the shedding of life-giving blood was required in order to set people free from sin's penalty. The complete verse we quoted at the beginning says, *"For the life of the flesh is in the blood, and I have given it to you on the altar to make atonement for your souls; for it is the blood by reason of the life that makes atonement"* (Leviticus 17:11).

But you see, an animal's blood could not fully cover the sin of a person. So God did something really amazing. He sent His Son, Jesus, from heaven to earth to be born and live as a human being on earth. Jesus lived out a sinless life, so He would be without blemish and could be the ultimate sacrifice for our sins. He literally shed His own blood when He suffered and died on the cross. The Bible tells us, *"God was also pleased to bring everything on earth and in heaven back to himself through Christ. He did this by making peace through Christ's blood sacrificed on the cross"* (Colossians 1:20, God's Word Translation). God made the payment for sin through Jesus' death on the cross and offers us the opportunity to receive His forgiveness.

The Bible tells us, *"For the wages of sin is death, but the free gift of God is eternal life in Christ Jesus our Lord"* (Romans 6:23). Have you received God's free gift? If not, you can right now through prayer. Tell God that you believe Jesus died on the cross to pay for all your sins. Ask God to forgive you for "missing the mark." Then, place all your trust in Christ and receive His forgiveness. If you have done this, God says you have passed from spiritual death into spiritual life. You are now a new creation in Christ – all because of the blood of Jesus! He says, *"Truly, truly, I say to you, he who hears My word, and believes Him who sent Me, has eternal life, and does not come into judgment, but has passed out of death into life"* (John 5:24). He also says, *"Therefore if anyone is in Christ, he is a new creature; the old things passed away; behold, new things have come"* (2 Corinthians 5:17).

So now you see how important blood is. It's important for you to have blood in your body so you can live on this earth, and it's also important for you to know that Jesus shed His blood to take away your sin so you can live eternally. Let's now spend the rest of this lesson learning about the blood that God designed to keep you alive and well.

Super Highway

We've discussed a lot of the body systems. Do you remember which ones? So far, we have discussed your skeletal system, muscular system, digestive system, renal system, and respiratory system. Do you know what links the organs in all those systems together? It's your **circulatory** (sur' kyoo luh' tor ee) **system**, which is what carries your blood! Think of it this way. Have you ever seen the conjunction of superhighways in a big city? In Atlanta, we have one that we call "Spaghetti Junction," because all the highways twist and turn around each other like cooked spaghetti. It is at this junction that you can go from the highway you are on to any of the other highways that travel around the giant city of Atlanta and out of Georgia to North Carolina, South Carolina, Florida, Alabama, Tennessee, or beyond. All these highways are connected to one another at this junction.

Your circulatory system is a lot like a highway system.

Well, your blood also travels highways and smaller roads inside your body. Your blood vessels are like a major road network. At the junction of your heart, all the vessels are connected to the different organs and body parts – from your brain to your toes and everywhere in between. It's your blood that travels about this network of vessels inside your body. Your blood cells are very much like trucks that transport supplies around the city.

Artery Highways and Capillary Byways

Let's take a look at the vessels that carry the blood through your body. There are three main kinds. The first kind, **arteries** (ar' tuh reez), carries blood away from your heart. Other blood vessels, called **veins** (vaynz), carry blood back to the heart. All the blood vessels near your heart are large. They are sort of like the largest six-lane highways found at the junction of a city. As they begin to work their way through your body, they branch out, each time getting smaller and smaller. Eventually, arteries become little arteries, called **arterioles** (ar tir' ee ohlz). Little veins, called **venules** (ven' yoolz), take the blood and join together to make the veins that bring the blood back to the heart.

Capillaries are the third main kind of blood vessel. They join the smallest arterioles to the smallest venules and have very thin walls. These thin-walled, tiny vessels are found all over the body, near the tissues. Capillary walls are so thin that they allow the blood to give its oxygen to cells and pick up carbon dioxide from cells. We call blood that is carrying oxygen to the cells **oxygenated** (ok' sih juh nay' ted) blood and blood that gave up some of its oxygen and is probably carrying a lot of carbon dioxide **deoxygenated** (dee ok' sih juh nay' ted) blood. This "deoxygenated blood" is still carrying oxygen, but because it has less oxygen than it did when it left the lungs, it's considered "deoxygenated." This means that oxygenated blood flowing through capillaries gives up some of its oxygen to the cells near the capillaries. It then picks up carbon dioxide. At that point, it is deoxygenated blood.

The arteries and veins run alongside each other throughout your body. Arteries have thick, strong walls. This is because they come directly from the heart and have a lot of pressure going through them as the heart pumps. The veins are thinner and don't have as much blood pressure, because they carry blood that is on its way back to the heart. Although your heart pumps the oxygenated blood to your organs, the deoxygenated blood in your veins is moved by your skeletal muscles squeezing the veins' walls to work the blood back up to the heart and lungs. You'll learn a lot more about this in the next lesson.

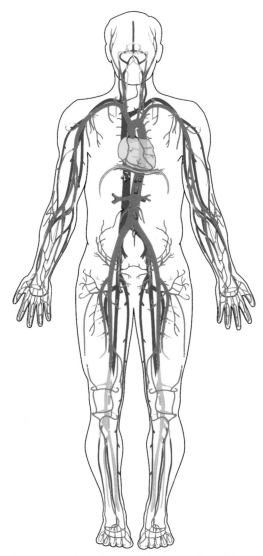

Arteries (red) and veins (blue) run alongside each other in your body.

Capillaries

Isn't it interesting that no matter where you cut yourself, you bleed? That's because there are capillaries under every single inch of your skin. If you've been an active child, you've probably experienced a number of injuries to different areas of your skin. If so, you may have noticed that certain parts of your body bleed more than others. For example, if you were to receive a small cut to your head, your blood would flow as if you had a bad wound. Yet, it takes a deep cut to your arm to bleed a great deal. This is because some areas of your body have many more capillaries than do others.

Capillaries are important because they are the vessels involved in providing your cells with the oxygen and nutrients they need to survive. They also pick up the waste products the cells must get rid of. The way this exchange occurs is very interesting. You see, the wall of a capillary is very thin – only one cell thick – so when the oxygenated blood travels through your capillaries near an organ whose cells need oxygen, the oxygen just seeps through the capillary walls and into the cells. In the same way, carbon dioxide from the cells just seeps through the capillary walls and into the blood. This seepage is called **diffusion** (dih fyoo' shun). Oxygen diffuses from where there is a lot of it (the red blood cells) to where there is little of it (the tissues). In the same way, carbon dioxide diffuses from where there is a lot (the tissues) to where there is little (the red blood cells).

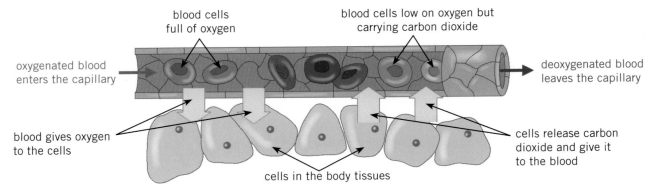

blood cells
full of oxygen

blood cells low on oxygen but
carrying carbon dioxide

oxygenated blood
enters the capillary

deoxygenated blood
leaves the capillary

blood gives oxygen
to the cells

cells release carbon
dioxide and give it
to the blood

cells in the body tissues

Here's something really interesting: though you can't grow new arteries or veins, you can grow new capillaries. In fact, this is exactly what happens when you work out your muscles. As the muscle is used, it needs more oxygen and nutrients than it used to. So it begins to grow more capillaries, allowing more blood to flow to that muscle. This makes your body even healthier, because you have better blood flow to your organs. Also, the more capillaries you have, the lower your blood pressure. It's like building more roads to a certain part of a city. These additional roads prevent traffic jams and congestion that occur when you have only a few roads leading into the city. That's why exercise, which helps grow capillaries, decreases your blood pressure. We'll talk more about blood pressure when we examine the heart in the next lesson.

Capillaries are also the connection points between your arteries and veins. When blood carrying oxygen gives it up and then receives carbon dioxide, it takes a different route home than the one it took to get there. So the blood begins on the "artery highway" to the organ, but after exchanging oxygen for carbon dioxide, it takes the "vein highway" back up to the heart and lungs.

Try This!

Would you like to see some real life venules, arterioles, and capillaries? Well, get a mirror and a flashlight. Hold the mirror up to your face. Now, shine the light towards the mirror. Open your mouth wide and position your tongue so that you can see the underside of it. The thick, blue lines are venules, the thick, pink lines are arterioles, and the thin lines are capillaries. You can also see them just below your eye if you pull the lower eyelid down and observe.

So now that you've learned about the different blood vessels, let's explore the blood that flows through them. Your blood performs four major functions inside your body. Let's study each of them.

Transporter

What if you showed up at the grocery store and there was no food there? Without the trucks that transport the food from the farmers to the store, you would be in big trouble. Transporters are essential. Well, your blood is also a transporter. After all, you just learned that it brings oxygen to your cells and picks up carbon dioxide, which it takes away from your cells.

Like a truck on a highway, your blood transports things all over your body.

In addition to transporting oxygen and carbon dioxide, blood also transports nutrients. Do you remember that when nutrients are consumed and moving along the gastrointestinal tract, they are broken down into smaller and smaller parts and then absorbed through the villi in the small intestine? Most of those nutrients are absorbed into the bloodstream. The nutrients are then delivered to the cells. Without your blood transporting oxygen and nutrients to your body, you couldn't do all the things you love to do, like running, jumping, and playing!

Protector

In addition to all the transporting trucks, you have some pretty powerful military vehicles traveling along your blood vessel roads! These military vehicles are there to protect your body from invaders and to speed healing. These vehicles in your blood are called **white blood cells**. They travel throughout your body, looking for any sign of foreign invaders, such as harmful bacteria or viruses. Should they find any foreign cells or other material (such as a splinter or a mean-looking cancer cell), the white blood cells will attempt to destroy it. We'll discuss this in more detail in the lesson on your immune system.

White blood cells are like tanks, protecting your body from invaders.

In addition to these military vehicles, your blood also carries **platelets** (playt' litz). These platelets jump into action when a blood vessel gets cut. A cut in a blood vessel is like a hole in a city wall. If the hole isn't patched up, evil invaders might sneak into the city. Also, when a blood vessel is cut, blood leaks out the vessel and can't continue its trip through the body. The platelets in your blood are a big part of the amazing blood clotting process, which seals cuts to keep your blood from leaking out and to prevent harmful bacteria from invading your body and making you sick.

Your blood carries messages throughout your body.

Message Carrier

If you've studied ancient history, you've probably heard about people who carried messages from one place to another. When a battle was won, for example, a messenger was sent to tell the king. If supplies were needed, a messenger carried the request for supplies to the place where the supplies could be found. When King Saul died, a messenger was sent to tell the future King David (2 Samuel 1:1-15). It didn't end up so well for the messenger, by the way.

Well, your blood contains many messengers called **hormones** (hor' mohnz). They rush about your blood vessel highways, carrying messages to different cells. These hormones are chemical messengers that help one part of the body know what another part is doing. When someone startles you, your body immediately produces a hormone called **epinephrine** (ep' uh nef' rin), which is also called **adrenaline** (uh dren' uh lin). This hormone tells your heart to beat more quickly and your breathing to speed up, in case you need to run. Your blood delivers many different hormones throughout your body so that normal body functions, like digestion, occur in a coordinated manner. You've probably noticed that your body has been growing. How much taller are you compared to this time last year? Interestingly, "growth hormone" travels through your blood, encouraging your cells to divide and make more cells, so that your body can get bigger. Without the blood to deliver this hormone all over your body, you would not grow!

Thermostat

Do you know the difference between a thermometer and a thermostat? A thermometer measures the current temperature, and a thermostat sets the current temperature. Your body's thermostat is in your brain, but one of the main ways your body temperature is controlled involves your blood. The fourth major function of your blood is to help control the temperature of your body.

Have you ever noticed that when a person is really hot, his skin gets flushed, turning his face more red than it normally is? This is because when your body gets too hot, it needs to release heat, so your blood will move more towards your skin, allowing the warm blood to release some heat into the environment. We'll discuss this more in the lesson on skin. Maybe you've noticed that when someone gets really cold, his lips turn blue. That's because when you get too cold, your body tries to keep as much warmth as possible by reducing the blood supply near the skin – especially the small blood vessels in your arms and legs. That's why people's hands and feet get frostbitten first when they are in extreme cold. It's also why you pull your arms close to your chest when you're cold. Your chest has a lot of blood flow and warmth, so it can help heat up your arms and hands. Exercise kicks those blood vessels into action. So, if you jog or do jumping jacks when you're feeling chilled on a cold day, some of that warm blood will come back to the surface and into your arms and legs, and you'll immediately feel warmer.

Before moving on, explain what you have learned so far, including the different kinds of blood vessels and the four major functions of blood.

Blood Basics

To you, blood may seem like just a red liquid, but it actually contains a lot of interesting stuff! Believe it or not, blood is actually a mixture of liquid things and solid things. Let's learn about the major components of your blood. To help you really remember what you are learning, you are going to create your own model of blood. Does that sound gross? It will be fun, I promise! Before you continue reading, gather the following things so you can build your model of blood as you learn about each component: A bowl, one cup of corn syrup, ¾ cup of candy red hots, one white jelly bean, and some candy sprinkles.

You will make a model of blood as you learn about it.

Plasma

Based on your experience with blood, it probably doesn't surprise you that there is a lot of liquid in your blood. The liquid is called **plasma** (plaz' muh), and the other components of your blood float in the plasma. To represent plasma in your blood model, pour corn syrup into your bowl. Plasma is actually straw colored, so the corn syrup is a good model for plasma, which makes up about 55% of the stuff we call blood.

The plasma itself is over 90% water. The other 10% is made up of little plasma proteins, tiny amounts of dissolved gasses (mostly oxygen and carbon dioxide), salts, vitamins, nutrients, hormones, and urea (a waste product that comes from protein breakdown). Much of the stuff that's transported around your body (besides oxygen) is transported by this liquid part of your blood.

Red Blood Cells

The next component of blood is the red blood cells, which make up about 40% of the stuff in a drop of blood. They are so abundant that they turn the plasma red! Pour your red hots into the bowl of plasma. Stir the red hots a bit so the entire mixture turns red. Red blood cells are more properly called **erythrocytes** (ih rith' ruh sytez). "Erythro" is Greek for "red," and "cyte" refers to a cell. Altogether, that's "red cell!"

You probably already know that erythrocytes transport oxygen. They are carried along through your blood vessels with the flow of the blood, delivering oxygen and picking up carbon dioxide as they go. Unlike the red hots, however, red blood cells are shaped like an inner tube or a lifesaver, with the center portion filled in. Each red blood cell is about 7 micrometers in diameter and 2 micrometers wide. For reference, a strand of hair is about 80 micrometers in diameter. As you can see, red blood cells are so tiny that they cannot be seen without a microscope. But because there are so very many of them in the body, they color the blood red.

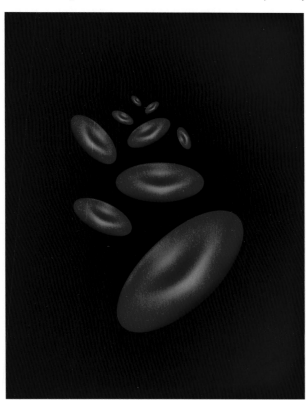

The special shape of red blood cells helps them to be very efficient at their job!

A red blood cell's special shape allows it to be more efficient at exchanging oxygen and carbon dioxide than it would be if it were round. It also allows the cell to be more flexible. This is good, since red blood cells spend their time rushing through crowded blood vessels that twist and turn. They can get through the blood vessels more easily, and they can last longer, if they are flexible. Some people have certain blood disorders that make their blood cells lose their special shape. As a result, the cells aren't nearly as good at carrying oxygen, so these people tend to tire out easily, since their tissues don't get enough oxygen.

Red blood cells can carry oxygen because of a special protein they contain called **hemoglobin** (he' muh gloh' bin). While this might sound like a frightening creature, it is just a carrier. It carries oxygen. How does it do this? Well, the protein has special centers that contain iron atoms, and oxygen is attracted to iron. When a red blood cell is near oxygen, the oxygen grabs on to the iron, and hemoglobin takes it for a ride in the bloodstream! Without hemoglobin, you would die of suffocation, because your cells would never get oxygen. Hemoglobin is the device God designed to carry oxygen throughout our bodies!

Another interesting thing about hemoglobin is that it gives your red blood cells (and your blood) their red color. When hemoglobin has oxygen attached to its iron atoms, it is bright red. When the oxygen leaves, it is still red, but it is dark red. This produces an interesting effect. Do you remember looking at your blood vessels in your tongue? We told you the blue lines were venules, which are the vessels your blood enters once it has given its oxygen to the tissues. Take a look at your wrist. The blood vessels that you see there are venules. They look blue as well, but it's not because the blood in them is blue. The blood in them has given up its oxygen, so it is dark red. Why do they look blue? Well, your skin distorts the color, turning the dark red into a blue. Even though they look blue, the blood running through them is dark red. If you have ever seen someone have blood drawn for a lab test or to donate blood, what color is the blood? It is dark red, because it is pulled from a vein that carries deoxygenated blood. So just remember that oxygenated blood is bright red, and deoxygenated blood is dark red.

Life of a Red Cell

You know a lot about the structure of cells, but red blood cells are different from other cells because they don't have a nucleus, nor do they have mitochondria. "How can that be?" you may wonder. Isn't it true that the nucleus contains the DNA, and mitochondria are the energy producers for the cells? Yes, but God designed red blood cells quite differently. You see, God needed the red blood cells to have a lot of space for hemoglobin. That leaves little room for those mighty mitochondria or the giant nucleus. In fact, each red blood cell contains approximately 280 *million* hemoglobin molecules. There's not much room for anything else. Unfortunately, this means the red blood cells aren't able to repair damage or make much energy. Because of this, they have a much shorter

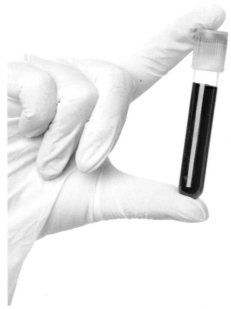

This deoxygenated blood was drawn from a vein, and it is dark red.

life span than other cells in the body. Most red blood cells live for only about four months. During every single second of your life, two *million* of your red blood cells die. Yikes! Thankfully, every second, your bone marrow produces two million new red blood cells to replace the ones that die.

Even though they live for only four months, red blood cells lead a very busy life for the short time they are alive. They are constantly rushing about your body, making deliveries. While they do this, they are forever bumping into one another and into the walls of your blood vessels. Have you ever bumped into a wall? It hurts, doesn't it? This relentless bumping roughens up the red blood cells' edges. When their edges get too rough, special cells called **phagocytes** (faj' uh sytez') show up and eat them ("phago" comes from a Greek word that means "eat")! Egads! I'm glad we don't have phagocytes walking around our house when we get a little worn out. After the phagocytes gobble up the worn-out cells, they head on over to the liver or to the spleen for final destruction.

In the liver, the iron is removed from the dead red blood cell, and your body sends it back to the bones to be used in the making of new red blood cells. God made your body a natural recycling factory. If God didn't create your body this way, you would need enormous amounts of iron in your diet just to survive. You see, when your body doesn't have enough iron, it can't produce enough red blood cells. Without enough red blood cells, you wouldn't be able to breathe or run around and play. We call this condition **anemia** (uh nee' mee uh), and it leads to feeling tired all the time. If the anemia becomes too severe, and oxygen delivery to the body is dangerously low, death can result. So go eat a piece of meat and a giant spinach salad so you'll have plenty of iron!

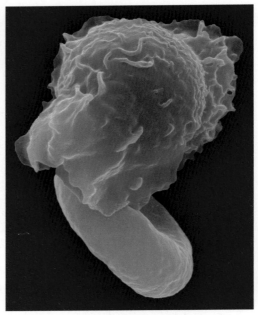

This microscopic image shows a phagocyte eating a worn-out red blood cell. The colors are not real; they have been added by a computer.

Try This!

Many foods (like cereal) are fortified with iron. You may think this is done with microscopic particles of iron, but it is simply flakes of iron poured into the box! Would you like to see? You'll need a box of iron-fortified cereal, a Ziploc bag, a strong magnet, and a mallet. Pour the cereal into the Ziploc bag and seal it (remove all the air before you seal it). Carefully crush the cereal with your hands, and then crush it even further by hammering it with the mallet. You need to be very careful so you don't break the bag or cause the bag to open and spill. Once the cereal is crushed, insert the magnet into the bag and move it around the contents of the bag. The iron shavings will stick to the magnet. Bring the magnet out and tap it onto a white piece of paper. Do you see the iron? This is what your body needs to create red blood cells!

Explain what you remember about red blood cells before moving on to learn about white blood cells.

White Blood Cells

White blood cells are called **leukocytes** (loo' koh sytez), because "leuko" means white. They are much less common than red blood cells. For every 700 red blood cells in your body, there's usually only one white blood cell. If you are sick, there will be a lot more white blood cells running around your body, because they are like little soldiers called into duty when there's an invasion in the land. They travel around, searching out germs and dead cell parts that can hurt you.

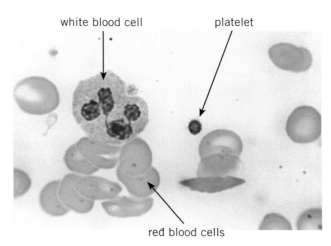

This microscopic image shows red and white blood cells and a blood platelet. The purple colors are not real; they come from stains.

Drop one white jelly bean into the bowl that holds your model of blood. It will represent a white blood cell. As you can see from your blood model, white blood cells are generally larger than red blood cells. As you already know, white blood cells play an important part in keeping you healthy. They fight infections and clean up debris and dying cells.

Although the contents of your blood are moved by the action of the heart pumping, some white cells can move on their own. They need to be able to do this because they have to locate and move to the things that need to be eliminated. They actually spend most of their lives in your tissues, not in your blood vessels. When they are called to duty, they squeeze in and out of the walls of your capillaries, going here and there to find infection or dead cells anywhere in your body. For example, white blood cells are found in your alveoli, where they eat any dust or germs that may have gotten into your lungs.

Platelets

Are you ready to add the last ingredient to your model of blood? Get out some sprinkles and drop about a teaspoon of them into your mixture. These represent blood platelets. Platelets are not cells; they are actually fragments of cells. Picture breaking a cookie up into smaller pieces and you'll have an idea about how platelets are formed. Large cells in the red bone marrow cast off fragments. These cell fragments circulate in the blood stream for about ten days, and then, if not used, they are removed by phagocytes.

By placing gauze over her wound, this child is helping her blood to clot so that the bleeding will stop.

Though they may seem insignificant among all those big blood cells, if you didn't have these little platelets in your blood, you would bleed to death whenever you cut yourself. You see, these platelets travel in the bloodstream until they detect damage to the blood vessel walls. Multiple platelets collect at the damage site, creating a structure that's like a logjam in a river. It helps to slow down the flow of blood through the damaged blood vessel wall. When clogging up the blood flow, the platelets send a message, calling special chemicals to the area. These chemicals cause the blood to clot, which seals the damage to the blood vessel wall so that no more blood leaks out. This all occurs within a few seconds of blood vessel damage – even damage as mild as a paper cut.

For small cuts, which only affect the tiny blood vessels close to the surface of the skin, it's easy for the blood clotting process to stop the blood flow. That's why a paper cut stops bleeding so quickly. After clogging up the flow with a scab, the injured area rebuilds itself over a few days or weeks, and the scab then falls off. When the damaged area is much larger, the blood clotting process can use a little help to stop the bleeding. This is why we apply pressure to wounds. With a clean cloth, we can use pressure at the damaged area to decrease the flow of blood, giving the clot a chance to form.

Wound Care

If you have a small cut or scrape, you should take care to treat it so that you do not get an infection.

1. Stop the bleeding with something clean. The best thing to use would be sterile gauze found in a first aid kit. Place the gauze on the injury and apply pressure.
2. Wash the injured area with soap and water and if you have an antibiotic ointment, put it on your cut before you put on a bandage.

If you have a large or deep cut that is bleeding heavily, seek medical help immediately. Also, if there is any foreign object sticking out of the injury, you should not try to remove it yourself.

1. Ask an adult to help you with your situation.
2. Apply pressure to the injury with a thick pad over a clean material. If possible, elevate the injured area to slow the bleeding.
3. If the injury is on someone else, try to avoid using your bare hands. Use latex gloves if available and if possible, the person who is injured should apply the pressure.
4. Have an adult take you to a medical facility where the medical staff can properly treat the injury.
5. If no adult is available to help you, call 9-1-1. The emergency staff on the line will get medical care to you.

If after any injury you develop a fever, or if your injury becomes red and swells or has red streaks around it, you need to call your doctor. This is a sign of infection.

Making Blood

Do you remember where your blood is made? Well, your blood cells are made in your bones! The red bone marrow inside your bones contains wonderful cells called stem cells. You'll hear that word more as you get older. Stem cells are truly amazing. In fact, they give us a glimpse of how creative God is. You see, a stem cell has the ability to become almost any kind of cell it needs to be. There are several different types of stem cells, but right now we'll only discuss the type found in bone marrow.

The stem cells found in bone marrow go through a series of steps to develop into the blood cells your body needs at any given time. Some of the stem cells in your bone marrow will develop into red blood cells. Some will start a process that ends up producing one of five different kinds of white blood cells. Still others will go on to make platelets. Isn't it amazing that one cell can develop into so many different things that your blood needs?

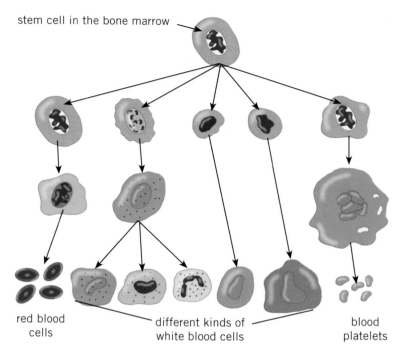

stem cell in the bone marrow

red blood cells

different kinds of white blood cells

blood platelets

Stems cells go through several stages before they become the different kinds of cells you will need in your body. This image shows how one stem cell can go through stages to become either red blood cells, white blood cells or platelets.

Need Blood?

So how much blood is in your body? Well, infants are born with about a liter of blood, which is a little over four cups. Most adults have about five liters of blood, which is about twenty-one cups. If you want to know how much blood you have in your body, take your weight and divide by 13, because about one-thirteenth of your body weight is blood.

As you can see, bigger people have more blood, while smaller people have less. That is why people that weigh less than 110 pounds cannot give blood. Why would anyone want to give blood? Well, when people have been injured and are losing a lot of blood, if they are not given more blood, they will die. Blood banks take blood from volunteers and store it so it can be given to those who have lost blood and need more. However, you can't just give anyone's blood to a bleeding person. You have to give the right type of blood. Let's learn about the different blood types.

Blood Types

Years ago, people learned that when a person loses too much blood, he or she dies. So they started transferring blood from healthy people to those who had suffered major blood loss. We call this a **blood transfusion** (tranz fyoo' shun). Sometimes the transfusion worked beautifully, but most of the time it resulted in more problems for the person receiving the blood. When scientists discovered that people have different blood types, they finally understood why transfusions were usually unsuccessful. Unless you give a person the right type of blood, it will harm him rather than help him! Now, doctors are able to give blood to people that will help them, because they understand what type of blood a given person needs.

Special markers, called **antigens** (an' tih junz), are attached to the surface of your red blood cells. They

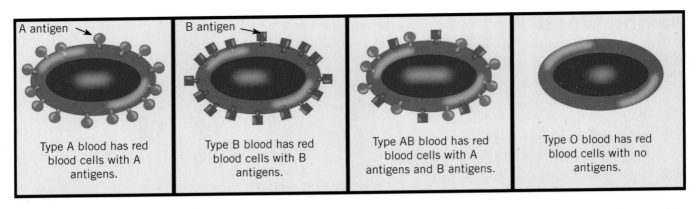

Type A blood has red blood cells with A antigens.

Type B blood has red blood cells with B antigens.

Type AB blood has red blood cells with A antigens and B antigens.

Type O blood has red blood cells with no antigens.

determine the type of blood you have. The four main blood types are **A, B, AB**, and **O**. These markers are like the flags on ships. They declare for all to see where the ship came from. If your marker (antigen) is an A, your blood type is A. If the antigen is a B, your blood type is B. If you have both A and B markers, your blood type is AB. If there are neither A nor B antigens on your red blood cells, you have type O blood.

If a marker (antigen) that isn't yours is found on a red blood cell, your body will destroy whatever cells are "flying the foreign flag." This can happen during a blood transfusion. If your blood is type A, your body knows that A markers are "you" and that they are safe. But your body will see B markers as enemies, and it will destroy cells flying the B flag. If you have type A blood and someone gives you type B blood, your body will note that the new blood cells do not belong. After all, they're flying an enemy flag. Those foreign cells will then be destroyed.

But, if you have type A blood and someone gives you type O blood, you will most likely be just fine. Type O blood cells don't fly any flags (remember – they have no antigens), so the body of the person with type A blood won't see the cells as foreign, because it won't see any enemy flags. As a result, it won't try to destroy them. For this reason, type O blood is called **universal donor blood**. It can be given to people with all four types of blood.

Now think about a person with type AB blood. Her body will recognize both the A and B antigens as normal, because her red blood cells fly both of those flags. No matter what flag a red blood cell flies, then, a person with AB blood will not see it as foreign. After all, that person's body recognizes A flags and B flags as belonging to itself. So people with type AB blood are called universal recipients – they can receive A blood because it flies the A flag, B blood because it flies the B flag, O blood because it flies no flags, or AB blood because it flies both flags.

So the problem with blood transfusions comes when you give a person blood that flies a *different* flag from what that person's blood flies. So you can't give AB blood to someone who has type A blood. The person's body sees the A flag as belonging there, but when it sees the B flag, it will attack immediately!

There's another marker on red blood cells that has to be considered, and it is the **Rhesus** (ree' sus) factor, or **Rh factor**. It was first discovered in Rhesus monkeys, so that's why we call

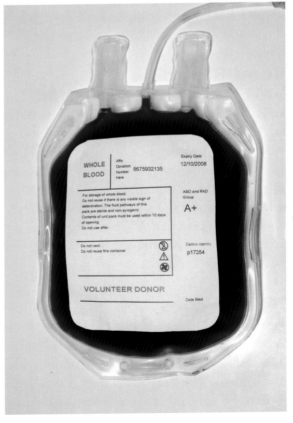

This is a picture of a bag full of donated blood. In order to help someone who needs blood, it must be compatible with that person's blood type.

it the Rhesus factor. Every person is either Rh-positive or Rh-negative. Rh-positive people have an Rh flag on their little red blood cell ships. Rh-negative people don't have any Rh flag flying at all. If you are Rh-positive, that means your body likes cells with or without Rh flags. So you can receive both Rh-positive and Rh-negative blood. But if your red cells are Rh-negative, that means you don't like Rh flags on blood cells. Your body will recognize red cells flying the Rh flag (Rh-positive cells) as foreign and will likely begin to destroy them. So, only Rh-negative blood can be given to Rh-negative people. But Rh-positive people can receive Rh-negative blood, just like people with type AB blood can receive type O blood.

When you talk about blood type, you should put the Rh factor and the letter together. A doctor might say that a man has "A positive" or "A+" blood. That doesn't mean he got a good grade on his blood test. It means his blood cells have the A antigen (they fly the A flag) and the Rh antigen (they also fly the Rh flag). A person with "O negative" or "O-" blood has no antigens on his blood cells at all. Remember, type "O" blood means the cells have neither A nor B antigens (they don't fly any flag), and "Rh-negative" means they don't have the Rh antigen either.

Every single day, blood is needed all over the world. People need blood when they are in a serious accident, or when they have to undergo a surgery that results in a great deal of blood loss. Sometimes people with blood disorders need to be given extra blood as well. So, having stored blood is very important. Many people donate blood through the American Red Cross. This blood is stored until it is needed. Sadly, there is almost always more blood needed than there is blood donated.

What Do You Remember?

Why did Jesus give Himself as the sacrifice for our sins? What does a person have to do to be forgiven for their sins once and for all and receive life everlasting? What are the four functions of blood? Name the four basic components of blood. What should you do if you are bleeding seriously? Where are your blood cells made? Why is it important to give people the right type of blood? What are the four blood types (not including the Rh factor)?

Notebooking Activities

Along with writing down all you learned about blood vessels and blood, make an illustration of blood along with each of its different components. Be sure to label each component and tell what it does.

Your final notebook assignment is to write out an **apologia** (uh pol' uh jee' uh). No, you're not going to say you're sorry! The word "*apologia*" is a Greek word that means "defense." When you tell others the reasons you believe in Jesus, you are defending your faith. (This is where the publisher of this book, Apologia Educational Ministries, Inc., gets its name.) Today you will create an apologia for your faith. Begin by explaining why God required animal sacrifices for sin. Then go on to explain what Jesus did to bring everlasting life to mankind. You can end with your personal experience as a Christian, if you like.

Experiment
Type Your Blood

A kit makes it easy to determine your blood type.

Can you see how important it is to know what type of blood you have? Believe it or not, you can buy an inexpensive kit to test your blood at home. You can find out how to order a blood typing kit by going to the course website we told you about in the introduction to the book. You will see lots of options. Once you determine your blood type, keep the information in a medical file that can be accessed easily if it is needed.

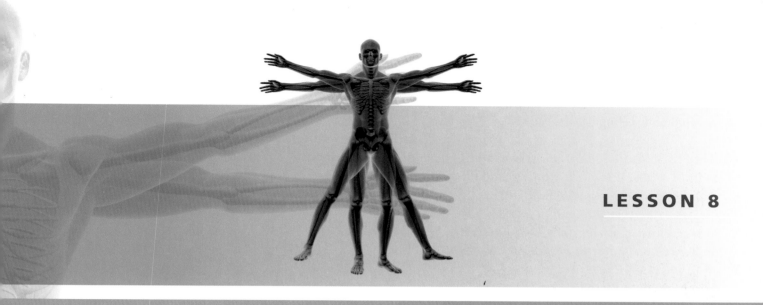

the CARDIOVASCULAR SYSTEM

It's time to move on to the heart of anatomy: I'm talking about your big, beautiful heart and all the vessels that carry your blood throughout your body! This is your cardiovascular system! The "cardio" part refers to the heart, and in anatomy, "vascular" refers to vessels that carry blood. The cardiovascular system refers to the heart and blood vessels in your body.

Today we know that the heart is a muscle that pumps your blood. However, the ancients believed that the heart was the place of a person's will and emotions and, in a sense, they were correct. In the Scriptures, the word "heart" often refers to a person's whole inner being – mind, will, emotions, affections, and desires. In fact, when God commands us to love Him, He asks us to do it with all our heart: "*You shall love the Lord your God with all your heart, and with all your soul, and with all your mind*" (Matthew 22:37).

References to the heart as the place of emotion and passion are common today. We talk about our *heartfelt beliefs*, and we enjoy having *heart to heart* talks with others. When we are really sad, we say we are *heartsick*; we might even be *heartbroken*. When someone is attractive to all those around, we might call that person a *heartthrob*. When your mom gives you a long lecture about your behavior, you really need to take it to heart. As you can see, people still think of the heart as the center of our emotions.

This painting by William Holman Hunt illustrates Christ knocking at the door of your heart, asking to be let in.

The apostle Paul tells us, "*that if you confess with your mouth Jesus as Lord, and believe in your heart that God raised Him from the dead, you will be saved*" (Romans 10:9). What does this mean? It means that your belief in Christ should come from your heart – the very core of your being; you should believe with conviction, passion, and feeling.

Ephesians 3:17 tells us that Christ dwells in the believer's heart. What does this mean? It simply means that His Spirit (the Holy Spirit) lives within us. But for Christ to really be "at home" in our hearts, we must ask Him to cleanse our hearts from sin and fill us with His Spirit each day. Jesus' desire to have continual fellowship with His people is revealed in this verse: "*Behold, I stand at the door and knock; if anyone hears My voice and opens the door, I will come in to him and will dine with him, and he with Me*" (Revelation 3:20). Do you invite Jesus into every area of your daily life? Your thoughts? Your emotions? Your hopes and dreams? Your worries? He desires to be at home in your heart. Will you let Him?

Heart Matters

Though the ancients understood that the heart was necessary for life, they knew little about what the heart actually did or how it worked. During the Renaissance period in history, when scientists began to understand how the different organs worked, the real function of the heart was discovered. Today scientists understand that the heart pumps blood throughout the body, allowing oxygenated blood to flow to the tissues to give them the oxygen they need. The deoxygenated blood is then carried back to the heart and then to the lungs, so it can pick up more oxygen and start the trip all over again.

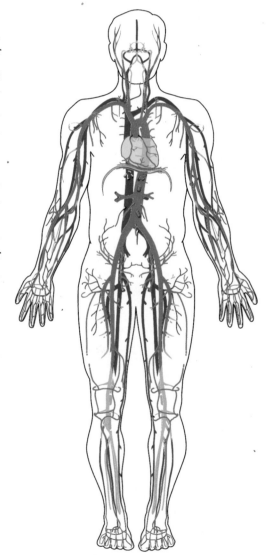

You may wonder why drawings of the cardiovascular system, like the one on the right, show some of the blood vessels in blue and some of them in red. You already know that blood is never blue. Why then do scientists represent some blood vessels as blue? They do this to illustrate the difference between the vessels that carry oxygenated blood and the vessels that carry deoxygenated blood. If you see a blue blood vessel in a drawing like this, you know that the blood inside the vessel is low on oxygen and must get to the lungs in order to pick up more.

Now let's take a closer look at the "cardio" part of the cardiovascular system: the heart. While we do this, we will build a model of the cardiovascular system with snack foods. Building the model should help you remember the important things we want you to learn about the heart and how blood flows through it.

Heart Anatomy

If you live in the United States, you may remember a time when you said the Pledge of Allegiance or stood to hear the national anthem. Most likely, you were taught to put your right hand over your heart in order to show resepect for your country. When you do this, you put your right hand over the left side of your chest. That's because the heart rests in the chest, slightly more on the left side than the right. There your heart sits, safely protected by your sternum and ribs in front and on the sides, and the spine in the back. The heart

Even though blood is never blue, red represents vessels that carry oxygenated blood, and blue represents vessels that carry deoxygenated blood.

actually sits between the lungs.

You may remember that the heart is mostly made of cardiac muscle, but there is a lot more to it than that! As you know, the main job of your heart is to pump blood throughout your body so the oxygen and nutrients carried in the blood can be used by the trillions of cells in your body – from your head to your toes and everywhere in between. Remember, oxygen is carried by the red blood cells, while nutrients are carried in the plasma – the liquid portion of the blood.

Make a fist. Now put it over the center of your chest. That's about how big your heart is and about where it is in your body. When looking at photos of a heart, it may seem like the heart is solid. Yet, inside it's a whole different story. In fact, it's two-stories high, like a two-story house! In this heart house, we call each room a chamber.

Each room is given a name. There are two rooms at the top that we call the **atria** (aye' tree uh). If you are referring to just one of them, you say **atrium** (aye' tree uhm). An atrium is the central hall of a home or building, and it's usually open to the sky above. In modern buildings, if there is a glass roof over the open area, we call that an atrium as well. The two atria in your heart are on either side, so they are called the **right atrium** and **left atrium**. Now remember, in anatomy, right and left refer to the person in whom the organ is located. In the drawing, then, the right atrium is on your left. The atria are the top rooms in the heart and open to enormous veins. Blood flows into the atria from these big veins.

The two lower rooms are called ventricles. There is one ventricle below each atrium. They are sort of like the basement. There are valves that open and close between the atria and ventricles, kind of like a basement door. So, essentially you have one atrium and one ventricle on the right side of your heart, and one atrium and one ventricle on the left side of your heart. The ventricles have thicker walls than the atria. Look at the drawing of the heart above. Do you see how thick the walls surrounding the ventricles are? Compare them to the walls surrounding the atria.

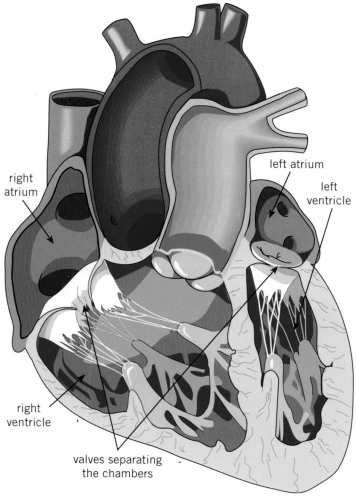

right atrium

left atrium

left ventricle

right ventricle

valves separating the chambers

This is a drawing of the inside of the heart, which consists of four chambers that are separated by valves.

Try This!

The design of the heart is amazingly complex. In fact, it is so complex that heart surgeons need at least seven years of additional training after they finish medical school just to be able to perform surgery on it! We are actually going to build a model of the heart; but since we don't have seven years to do it, we'll build a simplified one. The purpose of this project is to help you learn the names of some of the structures in the heart and to learn the path the blood takes through it.

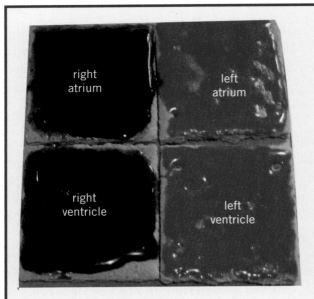

In this simple model of a heart, each graham cracker represents a chamber.

You will build your heart with graham crackers and colored frosting. If you don't have red and blue frosting, just use food coloring to color some vanilla frosting red and blue. Using one square of graham cracker for each chamber, frost the right atrium and ventricle with blue frosting, and then frost the left atrium and ventricle with red frosting. Remember, this won't look exactly like the heart, but it will help you understand the path the blood takes within your heart.

Pumping Iron

Surrounding your heart are several large blood vessels. We will create these for the heart model in a moment. These blood vessels are veins and arteries leading into and out of your heart. Remember, the arteries always take blood out of the heart, and the veins always bring blood into the heart. With one important exception, the arteries contain the oxygen-rich blood that will be distributed throughout your body. (If you want to know the one exception, keep reading!) Guess where the blood leaving your heart and heading out to your head, arms, and legs got that oxygen. From the lungs! So, the blood goes from the right side of the heart to the lungs, back to the left side of the heart, and then it is pumped from the left side of the heart throughout the body to provide oxygen and nutrients where they are needed. It drops off that oxygen, picks up carbon dioxide, and travels in veins back through the body to the right side of the heart. Then, it goes through the right side of the heart again, out to the lungs again, back to the left side of the heart again, and out to the rest of the body again.

You see, the heart is more than just a pump – it's actually two pumps working at once. The right side of the heart receives deoxygenated blood and then sends it to the lungs. That's why the two graham crackers on the right side of your model (your left, the model's right) are blue. The blue frosting reminds you that they deal with deoxygenated blood. The left side of the heart pumps the oxygenated blood out to the body, which is why the two crackers on the left of your model (your right, the model's left) are frosted red.

Now if this is confusing, don't worry. We will now lead you on a tour of what happens to the blood once it gets oxygen from the lungs. The bright red blood containing lots of oxygen leaves the lungs and enters the left side of the heart. It gets to the heart by traveling through four veins that come from the lungs. The veins carrying blood from the lungs to the heart are called **pulmonary** (pool' muh nair' ee) veins. They are the only veins in the body that carry oxygenated blood. All the other veins carry deoxygenated blood. Remember that veins always lead towards the heart, while arteries lead away from the heart. Now you know about the only veins carrying oxygenated blood.

Try This!

Let's add the four pulmonary veins to your heart model. Push some small marshmallows onto a toothpick and frost them with red frosting. Place them so they are attached to the left atrium. Remember, they should be red because they just came from the lungs and are full of oxygenated blood. These four pulmonary veins lead into the backside of the heart, directly into the left atrium. Look at the picture of a completed model on page 140 so you can see what this part of it should look like. So now you know why the heart is nestled snugly between the lungs! The heart gets all that bright red blood from both lungs, so it needs to be right in between them.

Now things get really interesting. You see, the atrium is a collection space. The blood from the lungs first fills the left atrium and then flows into the left ventricle where it's pushed out with a powerful force to the rest of the body. It leaves the left ventricle by moving through a large, important blood vessel called the **ascending aorta** (aye or' tuh). Look at the drawing to the right. The aorta is on top of the heart and is the largest artery in the body. It's about 1 inch wide. It has a thick, muscular wall so it can handle the high pressure of the blood leaving the left ventricle. The blood leaves with a great deal of force, because it has to travel all the way around your body.

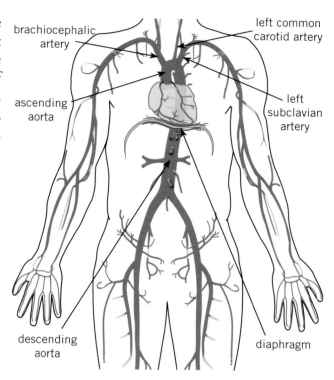

So, the ascending aorta is the main artery leaving the heart to deliver blood to the body. There are four vessels leading from the aorta. They become the major arteries in your body, and they lead to different parts of your body. The **largest artery** (the descending aorta) curves downward and goes towards your legs. The **brachiocephalic** (bray' kee oh suh fal' ik) **artery** takes blood to the upper right side of your body. The left common **carotid** (kuh rot' id) takes blood to your head, and the left **subclavian** (sub kaly' vee uhn) takes blood to the upper left side of your body. Find these major arteries on the diagram. Trace them with your finger. Now, see if you can figure out where they are on your own body. Those are some big blood vessels, aren't they?

Try This!

Let's add an aorta to our graham cracker heart. Put three large marshmallows on a toothpick and cover them with red frosting. That's your model of the aorta. Since your model isn't an exact copy of the heart, your marshmallow aorta will exit from the bottom of the heart. Remind yourself of the names of the chambers and whether the blood in them is oxygenated or deoxygenated.

The four large arteries you just traced branch out, becoming smaller arteries and eventually becoming arterioles. Do you remember that arterioles are little arteries? The arterioles divide and become capillaries. The capillaries have thin walls, allowing the oxygen and nutrients to flow out of the blood and to the cells. At the same time, the blood can pick up waste products like carbon dioxide from the cells. Capillaries then join together to form tiny veins called venules. Do you remember that when venules come together, they form veins? Eventually, veins come together to form larger and larger veins, in order to bring all that blood back to the heart.

Heart Health

You might think that your heart can get all the oxygen and nutrients it needs from the blood flowing within it, but that's not how it works. You see, the heart works so hard, and the heart muscle is so thick, that the heart must have blood vessels attached to its outside, penetrating through the muscle and supplying it with the energy it needs to function – just like every other organ and muscle in the body.

Do you remember from the previous lesson that you can actually grow more capillaries? People who work their heart with vigorous exercise have much stronger hearts because of all the capillaries that have grown in

137

and around the heart. Those capillaries provide more nutrients and oxygen and carry waste away with more speed. It's great for you to give your heart a workout every day!

People who are **sedentary** (sed' n tehr' ee), meaning they are inactive and do not get a lot of exercise, can have problems with their hearts and with the blood vessels near the heart. Other factors, including poor nutritional choices, smoking, emotional stress, and some diseases can also lead to heart problems. What happens is this: deposits of fatty material (called **plaque**) build up inside the blood vessels, including those supplying the heart muscle. This plaque makes it difficult for the blood to flow through the blood vessels, and the heart has to work harder. If the plaque builds up too much, the vessels get clogged, and the person can have a heart attack. Heart attacks are painful, because when the blood supply to part of the heart muscle stops, those muscle cells start to die from lack of oxygen. A heart attack is very serious and can lead to death if not treated quickly by a medical team.

Signs of a Heart Attack

Sadly, heart attacks are the leading cause of death in the U.S. for both men and women. It's important for all people to be informed about the signs of a heart attack, since it can happen at any time. Here are the signs of which you should be aware:

Chest discomfort: A tightening, pressure, squeezing, fullness, or painful feeling in the center of the chest that lasts for more than a few minutes. The discomfort can go away and come back. It could be just heartburn, but it's best to seek medical help in case it is a heart attack.

Upper body discomfort: A heart attack might also involve pain or discomfort in one or both arms, the back, neck, jaw, or stomach. Some people have been told that they are not having a heart attack unless their arm is in pain. However, the pain can be in other areas.

Difficulty taking a breath: Shortness of breath with or without chest discomfort may be a sign of a heart attack.

Other signs include breaking out in a cold sweat, nausea or lightheadedness. The earlier you treat a heart attack, the more likely it is that the person will survive it. If in doubt – get help!

Call 9-1-1 immediately if you think someone you know is having a heart attack. It's better to be safe than sorry. You might just save someone's life! When you are older, you can take a course in CPR (cardiopulmonary resuscitation) through the American Red Cross. Using CPR, you can help to keep the blood flowing through a person's body, even when the heart isn't working! CPR is useful in many situations; it's a great skill to have.

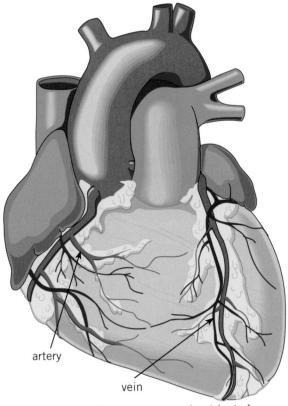

artery

vein

Cardiac muscle cells get oxygen and nutrients from capillaries that connect arterioles and venules, just like all the other tissues in your body.

CPR allows you to keep blood flowing in a person's body, even if his heart is not working.

Before we track the blood from the capillaries back up to the heart,
tell someone what you have learned so far in your own words.

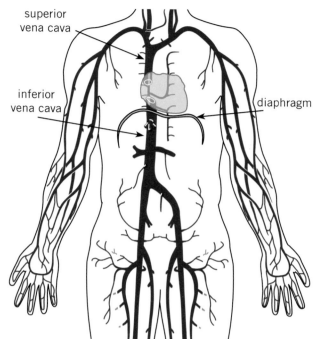

superior vena cava

inferior vena cava

diaphragm

Capillary Switch

We've followed the blood from the lungs, to the heart, through the aorta, through some major arteries, to the arterioles, and farther out to the capillaries. What happens now? Well, once the oxygen and nutrients are dropped off, the blood continues on a new track – sort of like how a train might move onto a new track to go a different direction. Your blood continues onto the vein track!

The capillaries give way to venules, then veins, then bigger veins, then bigger and bigger veins, until the blood has been collected into two main veins, the **inferior vena** (vee' nuh) **cava** (kay' vuh) and the **superior vena cava**. Inferior means below, and superior means above. The inferior vena cava collects the blood from the lower (inferior) parts of the body, while the superior vena cava collects the blood from the upper (superior) parts of the body. It might seem like the superior vena cava has an easy job, but remember what's in your head! That brain needs a lot of blood flow to keep you thinking clearly. Look at the diagram of the veins and trace your finger along veins that go into the inferior vena cava. Then trace all the veins that flow into the superior vena cava.

These two largest veins, the superior and inferior vena cava, dump deoxygenated blood into the right atrium. From the right atrium, the blood enters the right ventricle, where it is then pushed to the lungs through the **pulmonary arteries**. Do you remember that most arteries carry oxygenated blood? Well, these arteries are the one exception. They carry deoxygenated blood to the lungs so it can get more oxygen. And now we are back where we started, because once the deoxygenated blood picks up oxygen from the lungs (and drops off carbon dioxide), it gets pumped back to the left atrium so it can start the trip all over again!

So let's review how all this works. Oxygenated blood comes from the lungs through the pulmonary veins to your left atrium. It goes from the left atrium to the left ventricle, where it gets pushed into the ascending aorta and then takes one of four major arteries that begin branching into smaller arteries so the blood can reach all the body tissues. The arteries eventually become arterioles that take the blood to capillaries. When the blood is in the capillaries, it gives oxygen and nutrients to the cells and picks up waste products like carbon dioxide. It then enters venules, which lead to veins, which eventually lead to the right atrium of the heart. Blood then travels from the right atrium to the right ventricle, where it is pushed to the lungs through the pulmonary artery. Once it reaches the lungs, its oxygen supply is replenished; it drops off carbon dioxide, and then it heads back to the heart through the pulmonary veins.

Let's finish your heart model. Put large marshmallows on a toothpick and cover them in blue frosting. Add this "vein" to the upper blue graham cracker to symbolize the deoxygenated blood flowing into the right atrium from the inferior and superior vena cava.

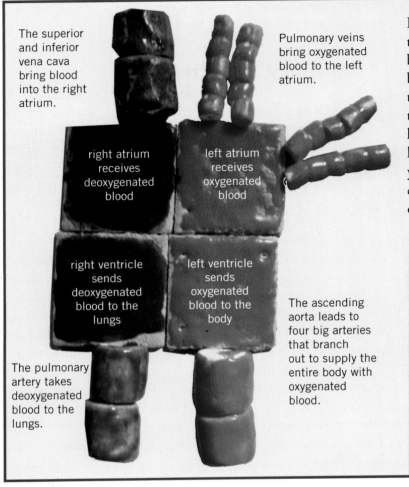

The superior and inferior vena cava bring blood into the right atrium.

right atrium receives deoxygenated blood

left atrium receives oxygenated blood

Pulmonary veins bring oxygenated blood to the left atrium.

right ventricle sends deoxygenated blood to the lungs

left ventricle sends oxygenated blood to the body

The ascending aorta leads to four big arteries that branch out to supply the entire body with oxygenated blood.

The pulmonary artery takes deoxygenated blood to the lungs.

Put large marshmallows on another toothpick and once again cover them with blue frosting. Add this "artery" to the bottom blue graham cracker to represent the pulmonary arteries leading away from the right ventricle. Congratulations! You have just made a simplified model of your heart! You might want to photograph it for your notebook. Remember, your graham cracker model is to help you learn the path of blood through the heart.

Show someone your graham cracker heart, naming all the parts. Explain why some parts are red and others are blue.

Cardiac Components

Remember that "cardio" refers to the heart. So, most of the names related to the heart will have "cardi" or "cardio" in them. The heart muscle itself is called the **myocardium** (my' oh kar' dee uhm). "Myo" means muscle, so you can guess what the word myocardium means. It means heart muscle! The myocardium is the thickest layer of the heart.

The heart sits inside a special sac that holds it in place, called the **pericardium** (pehr' ih kar' dee uhm). "Peri" usually refers to something that is on the outside; for example, the *peri*meter of a shape is the distance around the outside edge. So the pericardium is the surrounding outer area of the heart. This tough sac helps to protect the heart and keep it from moving around. The sac also keeps the large blood vessels attached to the heart. Without the pericardium, your heart might get tossed about a tiny bit when you jump on a trampoline, but God designed your marvellous heart to keep working, even if you have your pericardium removed!

The **endocardium** is the thin tissue that lines the inside walls of the heart. "Endo" means within or inside; so, endocardium means inside the heart. The endocardium is very smooth, and it continues into the blood vessels.

The thick, muscular wall between the two ventricles is called the **interventricular septum**. The thinner wall between the two atria is called the **interatrial septum**. Can you guess what septum means? It means a dividing wall and comes from a Latin word that means "fence." These walls prevent the mixing of the blood on the different sides of the heart.

Open Sesame

You know all about the atria and ventricles inside your heart, but you might wonder how your blood flows – and stops flowing – in and out of the chambers of your heart. This is done through the opening and closing of special little doors called valves. The heart has four valves. God placed a valve between the atrium and ventricle on each side. These are called the **atrioventricular** (ay' tree oh ven trik' yoo lur) **valves**. Study that word and see if you recognize any of its parts. Of course you saw parts of the words atrium and ventricle. So, it's easy to remember that the atrioventricular valves separate the atrium from the ventricle on each side of the heart.

You also have **semilunar** (sem' ee loo' nur) **valves**. These two valves each have three flaps that are shaped like a half-moon, which is why they are called semilunar ("lunar" means moon). One semilunar valve leads from the heart out to the aorta, and the other leads from the heart out to the pulmonary artery. The main job of each of these valves is to keep the blood from flowing backwards when the heart relaxes between beats. Do you remember where blood goes when it leaves the aorta? Do you remember where blood goes when it leaves the pulmonary artery? If not, you'll need to re-read the section above.

Try to recall all the heart parts about which you have learned, and then describe them to someone.

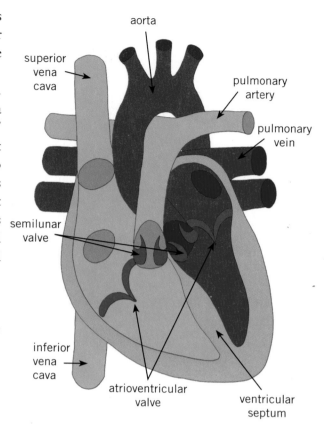

Try This!

To help you review the heart and how blood flows through it, we want you to color the heart outline you see on the right. Using the diagram above as a guide, color the chambers and vessels that have deoxygenated blood blue. Color the chambers and vessels that contain oxygenated blood red. Also, use arrows to indicate the path that blood takes as it travels into the heart, out the pulmonary arteries, back into the heart from the pulmonary veins, and out from the aorta. Finally, label the following parts:

- Right ventricle
- Right atrium
- Left ventricle
- Left atrium
- Inferior vena cava
- Aorta
- Pulmonary vein
- Pulmonary artery
- Superior vena cava

If you don't want to color and write in this book (and it is best not to), go to the course website that I told you about in the introduction to this book. There you will find the same drawing, which you can print out and color.

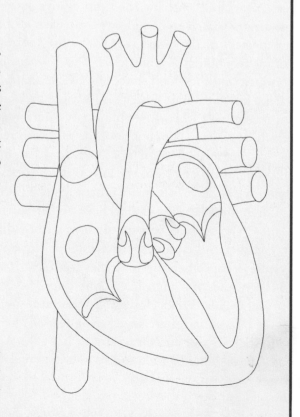

Beating Heart

Now you've learned all about the blood flow and some of the parts of the heart. But what is it that makes that beating sound? Let's find out! As you will learn in the upcoming lesson on the nervous system, muscles contract when nerves from the brain and spinal cord send a signal stimulating them to contract. Well, God, in His very great wisdom, set up a different plan for the heart muscle. The heart actually tells itself to beat. So, even if your brain is not functioning properly, your heart can still beat just perfectly. However,

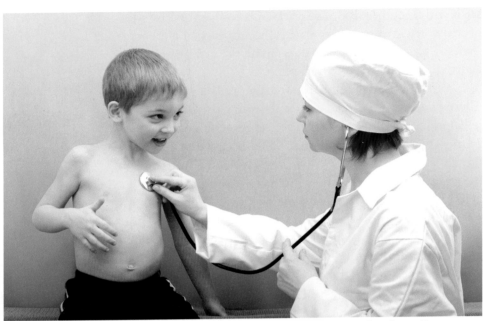

By listening to the sounds your heart makes as it beats, doctors can often learn if something is wrong with your heart.

your brain can affect how slowly or quickly your heart beats, increasing the beats if you are running, and decreasing them if you are relaxed and peaceful. But the heart beats on its own because it has something inside it called a **pacemaker**. This pacemaker actually tells the heart when to beat. So while all the other muscles in your body are stimulated by the brain or spinal cord, we say your heart is self-stimulating.

Let's go back to how the blood flows into the heart and explore what makes the beating we can hear and feel. The blood entering the heart always goes into an atrium, either the right or the left. This happens when the atria relax. Blood also flows from each atrium into the ventricles while the heart is relaxed. About seventy-five percent of the blood that enters the ventricles does so during this relaxed phase. Then the atria contract, completely filling the ventricles. This happens on both sides of the heart at the same time. Then the ventricles contract, and the blood starts to flow backwards, towards the atria. It catches in the flaps of the atrioventricular valves, however, and they close. Because they close, all the blood is forced out of the heart into an artery (either the ascending aorta or the pulmonary artery, depending on the ventricle). The pressure of the blood pushes the semilunar valves open, but once the ventricles relax, the semilunar valves close, making sure there is no backflow.

For a brief moment (usually less than half a second) all four chambers are completely relaxed. During this time, the blood from the lungs and the body again passes into the atria and ventricles. So, we are right back where we began. This entire process takes a little less than a second in most adults, but it takes even less time in children, because they tend to have higher heart rates.

The phase of relaxed filling is termed **diastole** (dye as' tuh lee). When the pacemaker triggers the atria to contract and complete the filling of the ventricles, the heart is in **atrial systole** (sis' tuh lee). **Ventricular systole** occurs when the ventricles contract. If you place your ear over someone's heart, you will hear two parts to every beat. It sounds like, "lubb dupp, lubb dupp, lubb dupp..." The first, deeper sound (the "lubb") is the atrioventricular valves closing. The second, shorter sound (the "dupp") is the semilunar valves closing.

These valves are an amazing feat of Divine engineering. Even though they make noise when they snap shut, they are soft and filmy, almost like silk. At the same time, however, they are incredibly strong. Only God could have designed such spectacular material! You will make a stethoscope at the end of this lesson so you can better hear your heartbeat.

Vascular Vehicles

You are a vascular person. This means you have a lot of vessels that carry fluid in your body. In fact, you have about 200 million feet of blood vessels in your body. Your mom has over 500 million feet of blood vessels in her body. That's about 100,000 miles! Can you guess which blood vessels you have the most of: arteries, arterioles, veins, venules, or capillaries? If you guessed capillaries, you guessed right! Why do you think God provided you with so many capillaries? Well, because they need to get close to the trillions of cells in your body in order for the blood to give them oxygen and nutrients. You actually have over one billion capillaries. Most cells are less than a millimeter away from a nearby capillary. Even as you grow, your body makes more and more capillaries to serve the new tissues.

Remember how the left ventricle pumps the blood with a powerful force out the ascending aorta? That force is great because it's got to move your blood throughout those miles of vessels. One hundred thousand miles is a long way to go! This rush of blood forced out of the heart with each beat causes the pulse.

Try This!

You can feel your own pulse by placing two of your fingers on the outside of the opposite wrist. Now, count the beats for fifteen seconds. Multiply that number by four, and you'll know how many times per minute your heart is beating! If you want a more accurate measure, count the beats for a longer period of time. For example, you could count the beats for 30 seconds and then multiply by 2. Alternatively, you can just count the beats you feel in an entire minute.

Now that you have felt your pulse, let's see if you can see your pulse! Place a small ball of clay directly over the spot where you felt your pulse. (This works best if you find the place where it is the strongest.) Now, stick a toothpick into the clay. What happens to the toothpick as your blood rushes past?

When you feel your pulse, you are feeling the "stretching out" of your artery caused by the blood being pushed from your heart. God designed your arteries to be elastic. The arteries' elastic tissue allows them to expand without popping when the pressure increases and then return rapidly to a smaller size when the pressure decreases.

Do you remember discussing **blood pressure** in the last lesson? Blood pressure is a measurement of the force of the blood pushing against the walls of the arteries. Although blood pressure is determined by cardiac output and other factors, some scientists believe that your our blood pressure will decrease if you have more capillaries for your blood to flow through. This is, in part, how exercise helps to lower blood pressure. The more you exercise, the more capillaries you form to feed the more active muscles, and the more often more of the capillary beds remain open. The pressure of your blood as it flows through your circulatory system depends on how many vessels it can flow through. The more vessels, the lower the pressure. As a result, blood pressure is lower when you have more blood vessels.

Blood pressure is an important thing to know, because high blood pressure can lead to all sorts of problems like heart attacks, while low blood pressure can lead to things like dizziness. Because of this, doctors measure their patients' blood pressure regularly. When they do this, they report blood pressure with two numbers. The first number indicates the pressure against the artery walls when the ventricles contract (**systolic blood pressure**). The second number is the pressure against the artery walls when the ventricles relax (**diastolic blood pressure**). The first number is always higher, because that's when your blood is getting its big push from your heart. When you were first born, your blood pressure was probably something like 90/60 (90 systolic and 60 diastolic). By the time you are an adult, your blood pressure should be about 120/80.

The less elastic your arteries are, the higher your blood pressure will be. Diets that are high in fats have been shown to stiffen the arteries, as have nicotine (from smoking) and stress. Continued exposure to high pressures can damage the delicate linings of the arteries, causing rough spots. These rough spots can cause blood clots. Blood clots are a good thing when a blood vessel has been damaged – they plug up the hole and keep your blood from leaking out. However, in a blood vessel that isn't damaged, the blood clot can plug up the vessel! This, of course, stops or slows down the blood flowing through the vessel, which means some cells won't get the oxygen and nutrients they need!

Before we close out this chapter, we will take a peek into the life of a single red blood cell named Zoe. We will follow her exciting journey through the human body! This will help you remember all that you've learned.

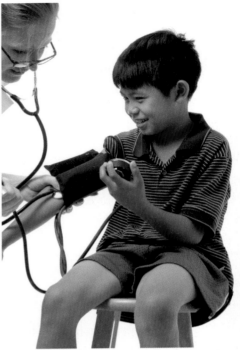

Doctors measure blood pressure regularly because it is one indication of your overall health.

Before we meet Zoe, explain what you have learned so far.

Zoe's Life

Zoe is a two-month-old red blood cell, coming back from the brain where she gave up some of her oxygen molecules. Many of the nutrients in the plasma she floats in were used up as well, because the person she lives inside is doing schoolwork, and the brain uses a lot of energy. She's bringing back with her some carbon dioxide, which the brain traded for oxygen. She's really hungry for more oxygen. It's been a long journey, even though it's been less than a minute since she left the heart. She's worried because some of the other blood cells jostled her, causing her to hit the sides of the wall of the artery she was traveling through. She thinks that she may have some rough spots on her, and knows this isn't a good thing! The more that happens, the sooner she'll have to meet with a phagocyte. But, she knows that whatever happens, she's living a productive life and doing great things for the person she resides in. She's doing what God meant for her to do!

Zoe enters the superior vena cava, flows into the right atrium, through the right atrioventricular valve, and down into the right ventricle. This is one of the most exciting moments of each trip through the body! Everyone gets ready. Suddenly the semilunar valve from the ventricle opens, and everyone is pumped out into the pulmonary artery towards the lungs by the force of the contracting right ventricle. Zoe is pushed into the right lung. She's happy about this, since the last time she took this trip, she ended up in the left lung. She enjoys the change in scenery!

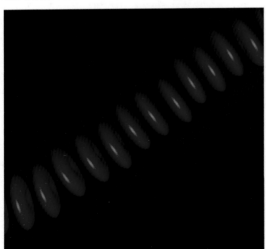

This image shows how red blood cells must stack together to get through a capillary.

As she travels through a pulmonary artery towards the right lung, the artery divides into smaller and smaller branches, each heading off in a different direction. Eventually, the artery gets so small that it becomes an arteriole. Zoe exchanges a few words with some friends and listens to gossip about white blood cells they saw in the toes where their person got a splinter this morning.

As Zoe travels through the arteriole, she knows that soon she's going to reach the capillary where she'll have to line up behind all the other red blood cells in front of her. She squeezes past a few other red blood cells, hoping to get in front. She looks ahead, and there before her is the small entrance hole to the capillary. Excitedly, she reaches the capillary and turns on her side to get through. Stacked up close on either side of her are her red blood cell friends. Moving along slowly, they pass by the alveoli. Here, Zoe deposits her load of carbon dioxide and picks up enough oxygen to fill each hemoglobin molecule she is carrying. Zoe is packed with hemoglobin, so she gets a lot of oxygen. She turns bright red as she fills her hemoglobin with oxygen. Now it's time to get that oxygen to the cells that need it!

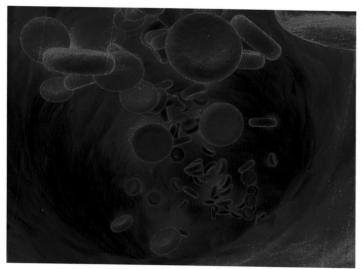

Unlike a capillary, an artery has plenty of room for many red blood cells.

She went to the brain last; where will she go this time? Some of the other blood cells are trying to guess where they may be going next. Zoe hopes it will be somewhere exciting! Moving along through the lungs, Zoe is thrilled when she finally enters the pulmonary vein and flows into the left atrium. She then flows into the left ventricle. The semilunar valve opens, and Zoe is pushed by the force of the left ventricle contracting up and out into the ascending aorta. "Wheeee!" she can hear some of the younger blood cells squealing with delight. She's two months old now. She no longer squeals on this roller coaster ride. That's for babies. But she wants to squeal, because it's a fast moving powerful ride out of the heart and through the arteries!

Soon Zoe is in a small artery and on her way to bring her treasure where it's needed. It seems things have changed. Her person is jumping up and down on a trampoline, which means her person is taking a break from schoolwork. Zoe gets excited, because she sees that she's headed to the leg muscles, which probably need a lot of oxygen right now! Down she goes to bring the quadriceps muscle oxygen so it can make the energy it needs to contract. In the leg, the arteries divide into smaller and smaller arterioles, so Zoe takes a sharp right and enters a capillary that feeds the right quadriceps muscle. One at a time, she and her friends stack up to drop off their oxygen and collect the carbon dioxide. As they do this, they take on a darker red color. Then, as she continues along with her carbon dioxide, she moves along the tiny capillaries as they come together to form a venule. The venule gets larger and turns into a vein, giving Zoe more room as she moves back up to the heart.

Zoe flows into the inferior vena cava and back into the right atrium again. She returns to the legs several more times, making new friends along the way. While traveling through the arteries a third time, Zoe learns that her person's breakfast has moved into the small intestine. So she figures she'll be heading to the intestines soon. There is never a dull moment for Zoe. She does this all day long. In fact, she will do this more than 1,440 times each day. She works very hard. No wonder she only lives for four months! Amazingly, it only takes one minute for Zoe to complete a round-trip journey beginning at the heart and returning back to the heart, and that's if her person is resting. If her person is exercising, Zoe can complete a round trip in as little as 15 seconds!

What Do You Remember?

What do we call the top two chambers of your heart? What do we call the bottom two chambers? What are the veins leading from the lungs into the heart called? What are the arteries leading from the heart to the lungs called? What is the main artery that takes blood out of the heart to the body? What are the names of the two veins that bring deoxygenated blood from the tissues of the body back to the heart? What is it that you are hearing when you hear your heart beat? What do the two numbers in a person's blood pressure mean?

Notebooking Activities

For your notebook, write an advertisement designed to make a stem cell want to become a red blood cell. In the advertisement, mention all the exciting places a red blood cell gets to go. Describe how the red blood cell will pass through the heart, including the order in which it passes through each chamber of the heart. After all, if you want a stem cell to become a red blood cell, it needs to have a good description of the job that it's going to do! If you ended up printing out and coloring the heart on page 141, include that in your notebook as well.

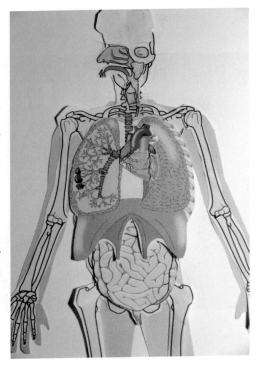

Personal Person Project

It is time to add a heart to your Personal Person. If you have the *Anatomy Notebooking Journal*, there is a drawing for you to cut out. If not, use the drawing on page 138 as a guide to make your own. Place it right in between the lungs, and remember: it takes up a bit more room on the left side, which is why the left lung is a bit smaller than the right lung.

Project
Make a Stethoscope

A stethoscope is a medical instrument that helps a doctor hear the sounds that are being made inside your body. This includes the sounds your heart makes as it beats and the sounds air makes as it passes through your respiratory system. You are going to make your own stethoscope so you can better hear what is going on in your own body.

You will need:

- A nine-inch balloon
- A small plastic funnel
- 18 inches of vinyl tubing (from a hardware store)
- Tape

1. Cut the neck off the balloon.
2. Stretch the balloon tightly over the wide opening of the funnel.
3. Insert the tubing into the small opening of the funnel.
4. Secure the tubing to the funnel with tape.
5. Place your stethoscope over the left side of your chest.
6. Can you hear the "lubb dupp" of your heartbeat?

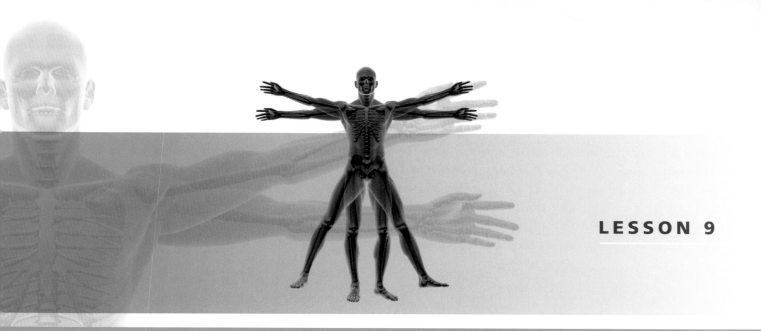

the
NERVOUS AND ENDOCRINE SYSTEMS

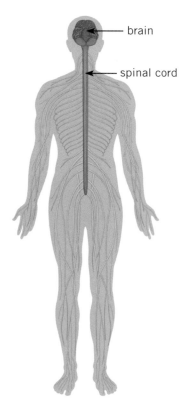

The brain and spinal cord (red) make up the central nervous system. All the other nerves (yellow-green) are part of the peripheral nervous system.

Have you ever wondered what causes your muscles to contract or how you can tell the difference between hot and cold? Do you ponder how you are able to feel love or perhaps why you can remember what happened last Saturday? How is it that you can even wonder about these things? Well, God gave you a special system called your nervous system that enables you to do all these things and much, much more. Without it, you really and truly wouldn't be you. Your nervous system controls a huge number of activities in your body, and scientists are discovering more of its wonders every day. We're going to explore the human nervous system, but don't be nervous! Your Creator has given you a brain capable of understanding and storing great amounts of information, a brain that quickly puts together new information, a brain that can gather incoming information and apply it to new situations. You can actually use your brain to understand your brain – what an amazing gift! And your fantastic brain is only one part of your nervous system!

In fact, there is so much to your nervous system that we will deal with it in two lessons. In this lesson, we're going to tell you some basic information about your nervous system. In the next lesson, we will be going into greater detail about how everything works.

The Central Highway

The nervous system is divided into two main parts, the **central nervous system** and the **peripheral** (puh rif' ur uhl) **nervous system**. The central nervous system has two parts. Only two. You'll need to know what these two parts are, so try to memorize this information. The two parts to the central nervous system are the **brain** and the **spinal cord**. That's not that hard to remember! These are really the most important parts of your nervous system. We sometimes call things that are most important "central." That's why we call the brain and spinal cord the central nervous system. We call the central nervous system the CNS for short. When you receive information, whether it's about how cold the pool is, how to do long division, or even whether or not you like broccoli, the information is processed in your brain.

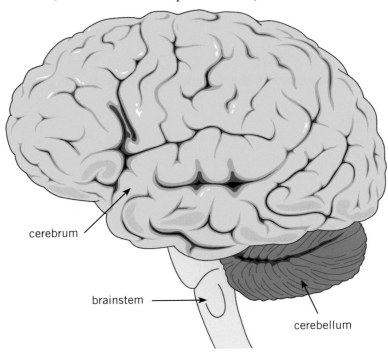

cerebrum

brainstem

cerebellum

The visible parts of the brain in this drawing are in different colors.

You can see three of the parts of the brain in the drawing on the left. The largest part, which is pink in the image, is the **cerebrum** (suh ree' bruhm). The cerebrum is where all your big thinking happens and also where most of your conscious actions are controlled. If your mom tells you to control yourself or gets upset that you can't remember your vocabulary words, tell her you'll speak to your cerebrum about it. We'll divide your cerebrum up into more parts in the next lesson.

Cerebellum (sehr uh bell' uhm) means little cerebrum. This is the little purple part of the drawing. The cerebellum is the part of the brain that controls the complex muscle movements you need to do, such as standing still without falling over. Even though you are standing still, lots of your muscles are relaxing and contracting in very small ways to keep you standing up. Your cerebellum takes care of all that, without you even being aware of it. When you learned to walk, your cerebellum got involved, helping your body coordinate all the muscle movements necessary to walk from one place to another without falling down.

The yellow part in the brain image is the **brainstem**. The brainstem connects the brain to the spinal cord so that all the rest of your body can communicate with your brain. It also controls some of the basic things you need to do to stay alive. For example, it controls your breathing. Do you remember the main muscle that controls your breathing? It's your diaphragm. Even though it is a voluntary muscle, you don't have to think about breathing, because your brainstem controls it. Now of course, if you want to control your diaphragm (to hold your breath, for example), you can use your cerebrum to "override" your brainstem's control of your breathing. However, if you aren't thinking about breathing, your brainstem is controlling it. It also does things like regulate your heartbeat and tell your salivary glands to make saliva when you have food in your mouth.

In the next lesson, we will get into more detail about the different parts of the brain. You'll learn where in the brain memory is stored, which part of the brain helps you see, which parts help you do math, and which parts give you the ability to draw and create art. But for now, we'll just focus on the three main parts: the cerebrum, cerebellum, and brainstem.

Try This!

To help you remember the different parts of the brain, we will build a model of the brain using clay. In the next lesson, your brain model will be even more detailed, but for now we'll keep it simple. You will want to use different colors of clay so you can easily identify the three different parts of the brain we have discussed. Begin by rolling a long piece of clay into a thin tube. Fold it over and over to form the lumpy, bumpy cerebrum. Next, form the cerebellum and place it under the cerebrum. Finally, form the brainstem and place it under the cerebellum.

Without the folds in your cerebral cortex, you might look something like this!

The outer part of your cerebrum (the visible part in the drawing on the previous page) is called the **cerebral** (suh ree' bruhl) **cortex**. You may be wondering why the cerebral cortex is so folded and bumpy. Can you think of a reason? Do you remember the little villi in the small intestine? What did they do? They gave the intestine more area to contact and absorb the nutrients being produced by digestion. Just as that design gives your small intestine more surface area, the folds in your cerebrum increase its surface area so that more information can be stored in a smaller space. God used the fold design quite a bit inside your body to make better use of space. This increase in surface area turns out to be very important! You see, the cerebral cortex contains cells that give, receive, and handle information. We call these cells **neurons** (nur' onz), and they are all over your nervous system. The greater the number of folds in your cerebral cortex, the greater the surface area, and therefore the greater the number of neurons you can have there. Without all those folds in your cerebral cortex, your brain would have to be three times bigger in order to do the same things it does right now! That would mean your cranium would have to be three times bigger! Someone might even mistake you for an alien!

Even now, as you're reading this book, your brain and spinal cord are receiving billions of messages – coordinating them, interpreting them, and choosing how best to respond to them. They're receiving signals from your skin and joints as to their locations and temperatures. They're receiving signals from your stomach about how the processing of breakfast is coming along; from your blood vessels about your blood pressure; from your back and leg muscles as they maintain your posture; from your eyes and ears as you interact with the environment around you. They're receiving signals from all over your body – about a million different messages every second! You didn't realize how busy you were, did you? I guess you can never again say that you're bored!

Can you imagine how distracting all this information would be if you actually had to pay attention to every detail of it, every second of every day? Thankfully, God designed the nervous system so that many of the incoming signals are not transmitted all the way to the brain. Many are filtered out before they reach the CNS. Others are handled unconsciously. This means that your brain takes care of them without alerting you. For example, without even thinking, you keep breathing in and out, in and out, thanks to your brainstem.

Peripheral Points

Now, let's chat about the **PNS** – that's the **peripheral nervous system**. The word peripheral is used to describe the outside part of something. For example, if you live near a big city but not in the big city, you could say you live on the peripheral edge of the city. The peripheral nervous system spreads out from the CNS, going to the outer edges of your body. Those things that spread out from the CNS are called nerves. Look at the drawing on the right to see how the nerves spread out to the outer reaches of your body. Do you see how the PNS is connected to the CNS? Why do you think that is? You'll understand in a bit.

The CNS connects to the PNS by 24 nerves that come directly from the brain (we call them **cranial nerves**) and 62 nerves that come from the spinal cord (we call them **spinal nerves**). These nerves come in pairs, so you have twelve cranial nerves going to one side of your head and twelve going to the other side. In the same way, you have 31 spinal nerves going to one side of your body, and 31 going to the other side.

Explain what you have learned about the CNS and PNS.

On My Nerves

I bet you've heard a person say, "He is getting on my nerves." But what is a nerve? Well, nerves are made up of cells called neurons. A **neuron** (nur' on) is the most important cell in the nervous system. Without your neurons working properly, you couldn't move, think, or feel anything. The word "neuron" comes from the Greek word "neuro," which means nerve.

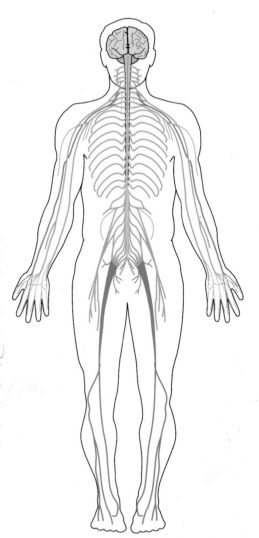

There are 31 nerves leading from each side of the spinal cord to the rest of the PNS. The 12 pairs of cranial nerves are not shown in this drawing.

Different neurons specialize in different activities, but they all have the same basic structure. Look at the drawing of the neuron to the left. Find a **dendrite** (den' dryt). Dendrites gather information. Do you see how they have arm-like structures reaching out in many different directions? These arms are like a bunch of roads leading into the center of a big city (the cell body). They reach outwards in order to funnel information towards the cell body. Why? Because the cell body contains the nucleus, which is the neuron's control center. Do you see the cell body and the nucleus? After the information is gathered in the cell body, it is

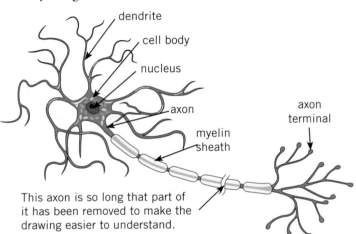

dendrite

cell body

nucleus

axon

axon terminal

myelin sheath

This axon is so long that part of it has been removed to make the drawing easier to understand.

Your brain and nerves contain millions upon millions of neurons.

sent along the **axon** (ak' son). Even the slowest axons transmit signals at about three feet per second. Some axons are wrapped by a **myelin** (my' uh lin) **sheath**, which is made of fatty tissue. This sheath allows information to travel more quickly down an axon. Not all axons have a myelin sheath. If an axon has one, information travels down it very quickly. The faster information travels in the body, the more efficient it is, so it is often good for axons to have myelin sheaths when you need information quickly.

Look at the axon again. Some axons are very short – just a millimeter or so long. Perhaps you already know how long a millimeter is. If not, get a ruler and find out. Other axons are very long. Can you make a guess how long the longest axon in your body is? Well, the axons that deliver information from your spinal cord to your big toe can be over a meter (about a yard) long. That's about the distance from a door knob to the floor!

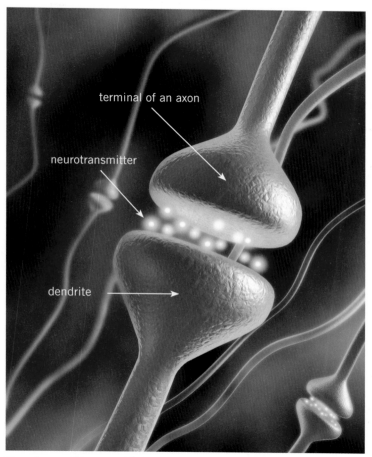

terminal of an axon

neurotransmitter

dendrite

Neurotransmitters send information across a synapse – a small gap between the axon terminal of one neuron and the dendrite of another neuron.

Now remember – the information a neuron receives travels down the dendrites to the cell body. It can then be sent down the axon in order to get it to the next neuron that needs it. Once the signal reaches the end (or terminal) of the axon, special chemicals are released. These special chemicals are called **neurotransmitters** (nur' oh trans' mit urz). If you look closely at the word neurotransmitters, you'll see how they got this name! They are chemicals that transmit information from one neuron to another. Neurotransmitters rapidly cross the very tiny space between the terminal of one neuron and the dendrite of another. This empty space is called the **synapse** (sin' aps). The empty space is very, very small. In fact, if you look at neurons through a microscope, they may seem to be touching one another, but they aren't. There is always a synapse between them. The synapse acts as sort of a "gatekeeper," letting necessary information cross to the next neuron, and stopping any unnecessary information that your body doesn't need at that moment. The neurotransmitters are sent across the synapse, and if the information is necessary, they pass it on to the next neuron. The receiving neuron gathers this information, and the entire cycle begins again. This happens until the information gets to its final destination. If you just slammed your finger in the door, the final location for that information is your brain. If you just hollered because you slammed your finger in the door, that holler started with a message in the finger that went to brain. Then another message came from the brain to the mouth and vocal cords. Lots of neurons were busy.

Each individual neuron is designed to send information in only one direction – from the dendrite, through the cell body, along the axon, and out to the terminals. So, some neurons send information from your body to your brain or spinal cord, and some send information from your brain or spinal cord to your body. We'll talk more about that later.

Because neurons receive and transmit a lot of information very quickly and make a lot of neurotransmitters, they use a lot of energy. Do you remember what mitochondria are? They are the powerhouses of the cell. Well, the cell body of a neuron has a lot of mitochondria! That's why your brain (which contains a lot of neurons) gets 20 percent of your body's blood flow, even though an adult brain weighs less than 3 pounds.

Try This!

Get some different clay with different colors, and make a model of a neuron. Make the dendrites and cell body out of one color of clay, and make the axon and terminals out of another color. Finally, add on a nucleus in a third color. Remind yourself which way the electrical impulse flows along the axon.

If you put a lot of neuron axons together and bundle them up, you'll get a **nerve**. Nerves are bundles of neuron axons running together. They gather information from inside and outside your body, and they send it to your brain. The information is processed, and messages are stored in your brain as memories. Some of this processed information can then be sent back to your body when needed. When you want to make a sandwich, or read a book, or perhaps play tennis, it's your nervous system that enables you to do it! This system enables you to plan the actions required and execute them in a precise sequence – so that the mayonnaise goes on your sandwich and not between the pages of this book!

Nerve messages from your brain and spinal cord control your muscles, both in your hands as you turn the pages of a book, and in your eyes as you scan the words. Nerve messages also help to control your rate of digestion and how much blood flow is going to your stomach. They even control the release of saliva into your mouth as you think about eating that mayonnaise-slathered sandwich!

Describe what you have learned about nerves to this point.

Sense and Do

Some nerves bring information to your brain, and others carry information away from your brain; however, most nerves take information in both directions. Do you think a single neuron can carry information in both directions? No. I mentioned before that each neuron can only carry information in one direction. But since nerves are made up of bundles of neuron axons, nerves can carry information both to and from the brain and spinal cord. In other words, within most nerves, there are neurons taking information to the brain as well as neurons bringing information from the brain. The neurons taking information to the brain are called **sensory neurons**. You have about 10 million sensory neurons in your body.

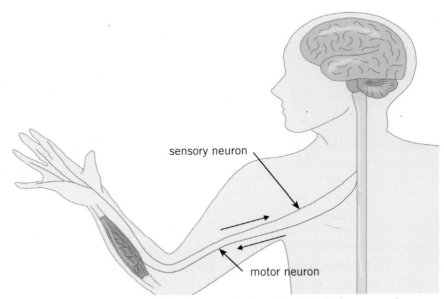

Sensory neurons send information to the spinal cord and brain for processing. Motor neurons send information from the brain and spinal cord to muscles and other organs to tell them what to do.

Can you think of some of the things you sense every day? Those things are sensed by your sensory neurons and sent to your brain.

The neurons taking information back to your body are called **motor neurons**. Think about all the parts of your body that you move every day. Each of these movements is controlled by many motor neurons acting together. You have about 500,000 motor neurons in your body. Think of it this way: sensory neurons sense and motor neurons do. You sense the stove is hot with your sensory neurons; you move your hand off the stove with your motor neurons.

Between it All

There is a third type of neuron, called an **interneuron**. Most of the neurons in your CNS are this kind of neuron. "Inter" means between, and the interneurons are found between other neurons. They connect the neurons! Sometimes they connect other interneurons. You have billions and billions of interneurons, and they are busily processing, sorting, and storing the information you are even now discovering!

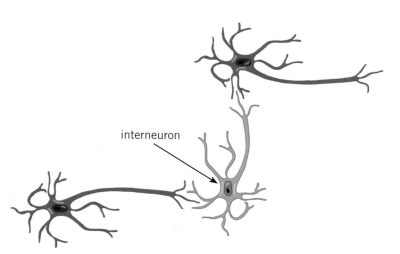

interneuron

Interneurons connect neurons together.

Integration Sensation

You may be wondering how information that enters your brain gets sorted and processed. Well, it happens by a function of your brain called **integration** (in' tih gray' shun). To integrate means to bring together lots of things into a combined, understandable whole. Let's say you are playing tennis, and you see the ball coming at you. Before you go to swing the racket, there's a lot of information that needs to come together in your mind, for example: the speed and location of the ball, the location in space of all the parts of your body, your plan about where you'd like the ball to land after you hit it, the force of the wind blowing, memories of hitting the ball before, and a million other factors! In a split second, all this information is integrated (brought together) in an organized fashion, so that you can (with practice) hit the ball over the net.

More specifically, your millions of sensory neurons bring much of this information to your brain. In your interneuron network, the incoming information is mixed with stored information and evaluated according to its importance. All of the important information is then integrated. Stored information might include: memories of ways you've hit the ball before, what happened when you hit it in certain ways, bits of past tennis matches you've seen, and the tennis instruction you have previously had. The result of this integration is sent out to your muscles from your brain. Do you remember which neurons send information from your brain to your muscles? Your motor neurons. What's amazing is that all this happens in a split second! God designed your brain to work with amazing speed. If you're having trouble processing information – like I was when I was on the tennis court, or my daughter is when she is trying to understand algebra – you might pray for better brain integration. That's where it all happens!

Sensing the SNS

Your peripheral nervous system is responsible when your pupils widen to let more light into your eye and when you wave at your friend. But these two things are controlled by different parts of your peripheral nervous system. Because of this, we divide the PNS into two separate systems. One is the **somatic** (soh mat' ik) **nervous system**, and the other is the **autonomic** (ah' tuh nom' ik) **nervous system**. Of course, we must shorten these long words to SNS and ANS.

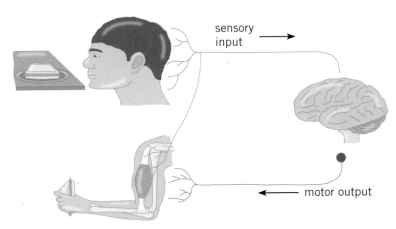

sensory input →

← motor output

Your SNS senses the sandwich on the plate and sends that information to the brain. You decide you want to eat it. The SNS sends information to the brain about the position of your arm, and then the brain sends instructions to the muscles in your arm to contract in just the right way so you lift the sandwich to your mouth.

You've already learned about your skeletal system and your muscular system, and you probably know a lot about your skin. You now know about sensory neurons and motor neurons. If you put this knowledge together, you've just discovered the somatic nervous system (SNS). You see, the SNS is responsible for the voluntary movements you make. It sends information from sensory neurons in places like the skin, muscles, and eyes to the brain so the brain knows about the outside world and your position in it. It also sends information from the brain to your muscles so that you can move them with purpose. When you see your friend, for example, the sensory neurons in your SNS send information from your eyes to your brain, and your brain sorts through its memories and recognizes your friend. It then sends instructions to your muscles to contract them in just the right way so you can wave to your friend.

So the SNS carries messages from your body to your CNS and then from your CNS to your body. The messages sent from your CNS are messages that you've given some thought to – even if you didn't notice thinking about them. So the SNS enables you to tell your fingers to turn the pages of a book, or your body parts to work all together to jump rope, or your hand to move the fork to your mouth. You're pretty much always aware of the somatic nervous system. You can tell yourself to lift your middle finger, and it will simply lift up. Do that now. Put your hands flat on a surface and lift your pinky finger. Now lift your thumb. Your SNS is working quite well, wouldn't you say?

Try This!

Let's try to trick your SNS. You will need a partner to do this activity. First, put your arms out in front of you. Then, cross them over one another. Turn your hands so that they face each other. Then, clasp your hands together. Next, roll your arms downward and inward towards your stomach, bringing them up close to your chest. Now have your partner point to, **but not touch**, one of your fingers. Try to move the finger your partner is pointing to. Can you do it? Have your partner point to a few other fingers. If you have trouble moving a finger when your partner points to it, it's because you have confused your SNS. It has lost track of which finger is in which position, so it has trouble knowing which muscle it needs to send commands to.

ANS Unaware

Unlike the somatic nervous system, the autonomic nervous system (ANS) works automatically. That's a good way to help you memorize which is the ANS and which is the SNS. Autonomic sounds like automatic. When we say something is automatic, we mean it works on its own.

The ANS functions without even letting you know what it's doing. It controls your smooth muscles (like the ones in your stomach) so organs can work without you thinking about them. Can you imagine if you had to sit down and concentrate, telling your stomach to grind up your breakfast, telling your pyloric sphincter to release the chyme into the small intestine, and so on, just to digest your breakfast? That would take all day! The autonomic nervous system takes care of all that without you thinking about it. In other words, it controls your involuntary muscles. You may hear your stomach rumbling when you're hungry, but you cannot consciously control the timing of your stomach contractions, because they are involuntary.

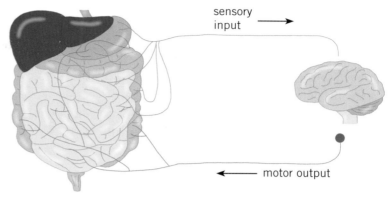

The ANS receives sensory information from the digestive system and based on that information, helps to control it automatically so you don't have to think about digesting your food.

Sometimes the ANS does involve your skeletal muscles and bones. Has anyone ever sneaked up behind you and yelled, "BOO?" Before you can even jump, your heart rate, respiratory rate, and blood pressure rise. Your pupils dilate to let in more of the available light, and your liver starts to pump out stored glucose. Blood is shunted away from "nonessential" activities like digestion (causing saliva secretion to decrease – that's why your mouth feels dry when you're frightened) and is sent to the skeletal muscles as your body tenses and prepares to run away or defend itself. Even though this reaction involves your muscular and skeletal systems, it is initiated by your autonomic nervous system, because it's not something you can consciously control. It's called the **fight or flight response**, and it's basically hardwired into your brain in order to keep you alive. If your brain had to consciously think about the initial response to a threat, you might not survive. But God in His wisdom has placed within us certain universal responses that enable us to protect ourselves from danger.

When you are frightened by something unexpected, your fight or flight response kicks in automatically.

Try This!

Because we're human, we all feel fear and anxiety at times. The next time you feel nervous, try to calm yourself by focusing on your heart rate for a few seconds. You'll notice that your heart feels like it's beating faster and harder than usual. Focus on taking steady, deep breaths while you tell Jesus about your fears. Then thank Him for His promises and faithfulness. After doing this for a while, pay attention to your heart rate again. What happened? Why do you think this happened?

Explain all that you know about the nervous system before moving on to the endocrine system.

Ending with Endocrine

Your nervous system controls a lot of what goes on in your body, but believe it or not, it is not the only body system that controls things. God designed a second system, the **endocrine** (en' duh krin) **system**, to control many activities that go on in your body. Instead of operating through neurons, however, the endocrine system works through chemicals called **hormones** (hor' mohnz). These hormones are chemicals produced in different glands found in your body and are sent into the bloodstream, where they interact with certain cells, telling them what to do.

Maybe you've heard of some of these hormones, like insulin. Insulin is produced by cells in the pancreas, and it tells cells to absorb sugar. If there is too much sugar in the blood, the pancreas releases insulin to get the cells to take in the sugar. So far, scientists have discovered about fifty hormones in your body. There may be many more!

Hormones are messengers. They move through your blood vessels, delivering messages – telling different cells and organs what to do and when to do it. Hormone means "to spur on," or "to set in motion," and that's what hormones do. For example, the **thyroid** (thy' royd) **gland** produces a hormone that speeds up ("spurs on") the rate at which almost all cells burn their fuel for energy. When the body needs more energy, the thyroid makes more of its hormone. When the body needs less energy, the thyroid produces less of its hormone. If your thyroid stopped working properly, your cells would not burn their fuel at the right rate. People with thyroid disease often gain or lose weight because the cells aren't using the fuel they have properly.

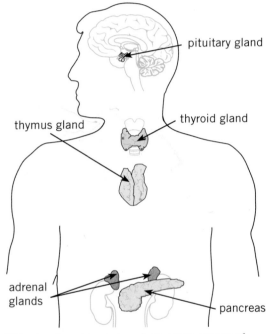

This drawing shows you where some organs of the endocrine system are in the body.

The central controller for many hormones is in the middle of your brain. It's called the **pituitary** (pih too' ih tehr' ee) **gland**. This gland is in your brain because it works closely with some of the neurons in your brain. So even though the endocrine system is a separate system from your nervous system, the two systems do work together.

Remember the fight or flight response? That's a good example of how the nervous system and endocrine system work together. When your ANS is jolted into action by fear, it triggers the release of multiple hormones into your bloodstream. One of those hormones is released by the **adrenal glands** that sit on top of your kidneys. This hormone, called **epinephrine** (ep' uh nef' rin), tells the cells in the liver to release any sugar they have stored up so that there is plenty of fuel in the bloodstream for those cells that have to work hard in running away or fighting.

Have you ever been frightened so badly that even after you realize everything is fine, your heart still races and you still feel jittery? That's because of the epinephrine that is still in your bloodstream. You see, when someone walks up behind you and yells, "BOO!" really loudly, your nervous system causes your muscles to jump, but it also causes your adrenal glands to release epinephrine. Once you realize there's nothing to be afraid of, your nervous system settles down, but the hormones are still flowing in your bloodstream. Until they are removed by your liver, they still stimulate various cells to get ready for the fight or flight response. As a result, you feel jittery until the epinephrine is finally gone from your bloodstream.

While many of the hormones in the endocrine system affect a large number of cells in the body, some hormones are more targeted. For example, the **thymus gland** makes a hormone specifically targeted to certain stem cells. These hormones cause those stem cells to develop into a certain type of white blood cell. This helps the body control its level of defense against disease.

In the end, the nervous and endocrine systems work together to help your body achieve **homeostasis** (ho' mee oh stay' sis). Homeostasis occurs when all the systems of your body are working together to maintain a stable, healthy condition. These two systems could not work alone. Millions of times throughout each day, your nervous and endocrine systems are sending and receiving messages that work to benefit your entire body. This is an amazing design from our amazing Creator!

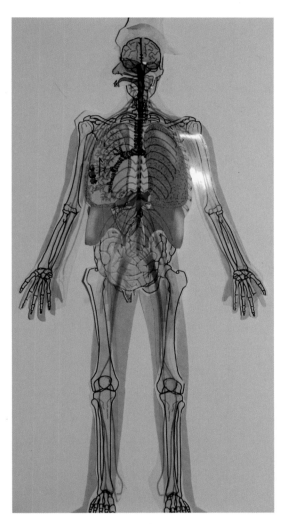

What Do You Remember?

Name the three parts of the brain we discussed. What part of a neuron receives information? What part sends the information on its way? Where do you find the nucleus in a neuron? What does a sensory neuron do? What does a motor neuron do? What is the somatic nervous system? What is the autonomic nervous system? What does the endocrine system do?

Personal Person Project

Don't you think it's about time your Personal Person had a brain? That head's been empty long enough! Up until now, your Personal Person has been much like the scarecrow from the *Wizard of Oz*, having no brain. If you have the *Anatomy Notebooking Journal*, there is a brain (and nerves) for you to cut out. If not, draw the CNS and PNS based on the drawing on page 152. Place the brain under your Personal Person's head so the nerves run throughout the body.

Notebooking Activity

For your notebook, draw and label the three parts of the brain we discussed. Also, draw a neuron and label all its parts. Write down everything you remember from the lesson! After that, create a Venn diagram like the one below to compare the central and peripheral nervous systems.

In this space, write things that are true about the CNS but not the PNS.

In this space, write things that are true about both the CNS and the PNS.

In this space, write things that are true about the PNS but not the CNS.

A Venn diagram uses overlapping ovals to compare things.

Project
Anatomy Trivia Game

We are going to use this lesson's project to really work your brain, because it is designed to help you remember all you have learned in anatomy and physiology so far. You will create a board game with questions from all the other lessons. You will assign points for each question, fewer points for easy questions and more points for harder questions.

You will need:

- Different colored pieces of paper on which you will write down your questions. Each color will represent a different body system.
- A file folder to create your game board
- Colored markers to draw your game board
- Game pieces made out of self-hardening clay in the shape of people or body parts
- Dice

1. Go through the book and write down questions that you want to include in your game. Assign points to each question based on its difficulty. Easy questions should be worth one point, harder questions should be worth three points, and the really hard ones should be worth five.
2. Assign a different color to each system of the body you have studied so far.
3. Put each question on a card that has the appropriate color for the body system the question is based on. As you work through the rest of the book, you can add to your game!
4. Draw your game board with the different colored squares, using the colors you chose for the body systems.

Here is how the game is played:

1. The first player rolls the dice and moves forward the number of squares shown on the dice.
2. After landing on a square, the person must pull a card that matches the square's color and answer the question.
3. To receive points, the question must be answered correctly. Record the points earned and tally them as you go along.
4. Players continue to move around the game board until a player reaches the last square.
5. Once one player reaches the last square, everyone else gets one more turn. That way, everyone has answered the same number of questions.
6. Whoever has the most points at this time wins!

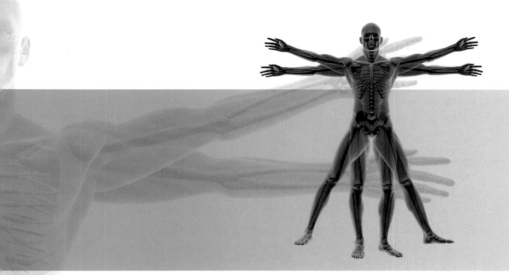

the
NERVOUS SYSTEM
EXTENDED

You probably remember that the central nervous system is made up of the brain and the spinal cord. You know that most of the cells that make up the human body are very, very small. But think about this: the brain contains about 100 billion neurons and many, many more supporting cells. That's what it takes to make you as smart and skilled as you are!

Yet with all these neurons in the brain, you'd think everyone would be a genius. But scientists have discovered that intelligence depends more on whether or not your interneurons are connecting a lot of neurons to other neurons, not on the actual number of neurons you have in your brain. The more connected your neurons are, the more intelligent you are. It's not about a bigger brain; it's about making connections from one neuron to another.

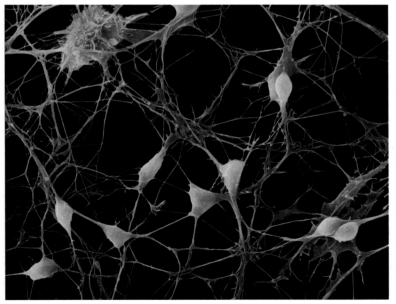

The connections between the neurons you see in this artificially-colored microscopic image are more important in determining intelligence than the actual number of neurons. The green cell at the top is not a neuron. It is a cell that helps to support the functions of neurons.

Try This!

Get a scrap of paper (about 4 inches square) and put dots all over the page a few centimeters apart. Then with a different color pen or pencil, make lines to connect the dots. Try to connect every dot to every other dot. Do you see how hard it is to connect each dot to every other dot? That gives you an idea of the incredible connections that are going on right now in your brain.

How are these connections in the brain made? Well, interneurons connect neurons to one another. They connect through learning and experiences. Your interneurons form better connections the more times you hear something, the more often you do something, and the more you experience something. That's why you are learning so much in science this year. You see, you start with reading the material, then you retell it, next you write and draw about it, and finally you do an experiment on it. You have been busy forming interneuron connections, and you didn't even know it! In reality, you're getting smarter just by doing this course. Studies have found that if older people will continue to pursue active learning, they'll keep their brains from deteriorating. You can continue to make neural connections, even when you're 100 years old!

Half a Brain

Do you remember the three parts of the brain you studied in the last lesson? The cerebrum, the cerebellum and the brainstem. Well, let's take a closer look at these three parts. If you forgot which part was which, go back and study that again. Remember, the more you are exposed to this information, the more likely neural connections will be made so you can remember it.

When scientists first started studying the brain, they realized that the brain is made up of two halves. Both halves look very similar. These two halves are named the **right hemisphere** and **left hemisphere**. So the cerebrum has two hemispheres, which are pretty similar in shape. These two halves are connected to one another by nerves, but they operate separately. Let's see how each part works.

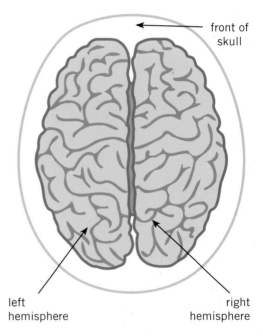

front of skull

left hemisphere

right hemisphere

This view of the top of the brain shows that the cerebrum is split into two hemispheres.

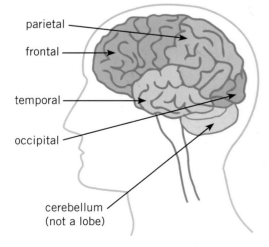

parietal

frontal

temporal

occipital

cerebellum (not a lobe)

In this drawing, the different lobes of the cerebrum are given different colors.

Shapely Cerebrum

The cerebrum is divided into different areas, which we call lobes. The lobes are: the **frontal** lobe, the **parietal** (puh rye' ih tuhl) **lobe**, the **occipital** (ok sip' ih tuhl) **lobe**, and the **temporal** (tem' pur uhl) **lobe**. Look for the frontal lobe in the drawing on the left; it is in the front of the brain. The parietal lobe is right behind it. The temporal lobe is about where your temple is. The occipital lobe is in the back. We'll build another brain model in a moment to help you remember all the lobes of the cerebrum. In the meantime, you will need to draw or print a picture of a brain to use while you learn about these different lobes. If you have the *Anatomy Notebooking Journal*, there is a page provided for the drawing.

Scientists have used brain scans to learn which parts of the brain are used during certain activities. These scans light up certain parts of the brain when different activities are performed. By looking at brain scans that are taken while people are doing certain tasks, scientists can learn which part of the brain controls which kind of task.

Frontal Fractions

On your illustration of the brain, draw the frontal lobe. Then, write down which activities are performed there as you learn about them. The frontal lobe is where speech and language is learned and used (write speech and language on your paper). It is also where your thoughts and ideas develop. This area of your brain is where you process your emotions. Skilled movement – like doing needlepoint, ballet, or brain surgery – is developed in this region as well. The frontal lobe processes your sense of smell, as does the temporal lobe.

Temporal Tones

The temporal lobe (draw this region on your paper) recognizes the tones and loudness of sound and is very active in the storing of memories. So when you listen to (and try to memorize) a new piece of music, you're using your temporal lobe. As mentioned above, the temporal lobe is active in processing your sense of smell. Have you ever smelled something that reminded you of somebody? Has a certain odor ever prompted a memory of some place you've visited? Well, because the temporal lobe processes both smells and memory, it's not surprising that smells are closely related to memories.

Occipital Optics

When something flies by your face and barely "catches your eye," or when you're staring at a crackling fire, you're using your occipital lobe. This lobe processes visual information.

Parietal Position

The parietal lobe responds to sensory information such as temperature, pressure, touch, and pain. Do you remember learning about how the brain integrates information? Well, the parietal lobe takes all sorts of sensory information and integrates it to make sense of the outside world and your position in it.

Now, if you've ever thought about it, you may have realized that because the brain has two hemispheres, there are two of each of these lobes – one in the right hemisphere and one in the left hemisphere. But the lobes don't simply duplicate one another; instead, they complement each other. Each side of the brain has different strengths. For instance, math, logic, and language are thought of as left-brained activities, because when people solve math problems while having their brain scanned, the left side of their brain lights up more than the right. The right side is more active with creative pursuits, like art and music. It also tends to be more active when you are dealing with shapes and images. It also is the most involved in helping us recognize faces we have seen before.

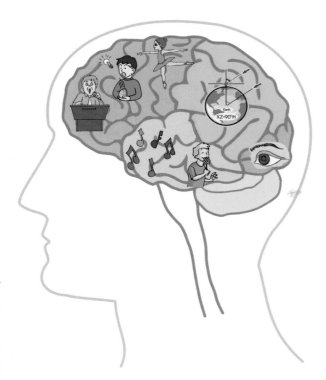

Even though the entire brain works together, each lobe concentrates on certain kinds of tasks.

Try This!

Just as people are either right-handed or left-handed, studies show that people can be more skilled at using the right side of their brain over the left, and vice versa. We describe people as being "left-brained" or "right-brained," depending on which side they tend to use the most. There are fun tests you can take on the Internet to determine whether you are left-brained or right-brained. Go to the course website we told you about in the introduction of this book. There you will find links to websites that have tests which will tell you whether you prefer using the right side or the left side of your brain.

One of my children is very right-brained, which explains how he can create such amazing drawings and make up music on his guitar. Another one of my children is very left-brained, which explains why school is so easy for him. When I took the test, I turned out to be equally left- and right-brained. Nevertheless, whether you're more left-brained or right-brained, the two sides are connected. Because of this, you can use both halves of your brain at any time. Even if one side is used more often, you can still use the other side very well.

Try This!

Do you remember how the connections between interneurons are formed? The more times your brain is exposed to information, and the more ways it is exposed to that information, the easier it is to remember the information later. That's because different and varied exposures to information help form different interneuron connections, increasing a person's chances of really retaining that information.

So get out your clay, and let's build a model of one hemisphere of the brain. As you add each part, remind yourself what that part does. Start with the frontal lobe, and then add the parietal lobe, the temporal lobe, the occipital lobe, the cerebellum, and the brainstem. If you say aloud the names of the lobes and their functions as you add them, they'll be even easier to remember later on!

What's the Matter

Another thing that scientists noticed when they first looked closely at the brain was that the outer surface is grayish brown, while the inner part is pinkish white. Keeping things simple, they called the grayish part **gray matter**, and the whiter part **white matter**. The neuron cell bodies can be found in the gray matter. In fact, it's the cell bodies that make the gray matter gray. The gray matter is found on the surface of the brain. Do you remember that the surface of the cerebrum is called the cerebral cortex? Ah! We're making lots of neural connections today! The cerebral cortex is only about 2-6 millimeters thick. That's not very thick!

Try This!

Stack four pennies on top of one another. Your stack is about 6 millimeters in height. That's how deep the gray matter of your cerebrum is!

My Myelin

Do you remember learning about myelin? It's the fatty insulation that wraps around the axons of your neurons. Do you remember what myelin does? It makes the signal going down the axon travel faster. That's good, because it means the signal will reach its destination sooner. Well, turns out there's lots of myelin wrapped around the axons of the neurons in your brain. The white matter contains the myelinated axons. It's the myelin that makes the white matter white. The more myelin you have, the more white matter you have.

Do you remember that a nerve is a large bundle of neuron axons? That's true in the peripheral nervous system, but in the brain, the same bundles are called "tracts," not nerves. Tracts in your brain are a lot like phone lines, connecting parts of your brain with other parts.

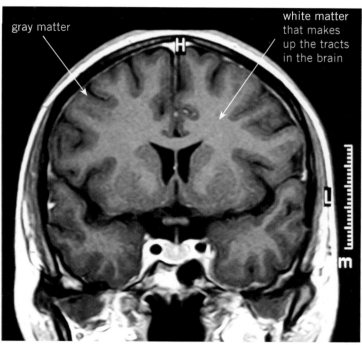

gray matter

white matter that makes up the tracts in the brain

This is an image of a brain as if it had been sliced down the center from ear to ear. It was done with an MRI – a machine that allows you to make such images without ever touching or seeing the brain!

Try This!

How well are you interconnected and myelinated? Let's do a fun activity to see how quick your reaction time is. You will need a ruler and a partner. Hold out your hand as if you are going to pinch something between your thumb and fingers. Have your partner hold the ruler vertically between your thumb and fingers. The numbers should be increasing the farther up they are from your fingers, and your partner should line up the 1-inch mark with your topmost finger.

Have your partner drop the ruler. Your eye will see the ruler drop then relay that information to your occipital lobe. Your occipital lobe will relay a message to your frontal lobe, which will send a message down your spinal cord to your hand. Finally, your fingers will contract to catch the ruler. That's a lot of stuff that has to happen! Let's just see how fast it happens!

After you react and catch the ruler, write down the number on the ruler that is closest to your topmost finger. The larger the number, the longer it took you to react to the ruler being dropped. Do this a few more times to see if you can decrease your reaction time. Now turn it around and have your partner be the one to catch the ruler. Is your partner faster or slower than you?

Before moving on to the cerebellum, explain what you've learned about the cerebrum.

Swinging Cerebellum

What if every time you looked to your left, you fell to the left? Or if you raised your hand, you fell backwards? Well, this is exactly what you did when you were a baby. Thanks to your cerebellum, you don't do that anymore! In the cerebellum, you can stand up without falling over. You can walk without wobbling back and forth. The cerebellum allows you to swing on a swing, jump on the trampoline, and dance with smooth movements. Although your cerebellum can, and usually does, integrate visual information when it's keeping you balanced, the visual input isn't usually needed.

Try This! Close your eyes and fully extend your arms out to your sides. Now, point outwards. Take the tip of your right index (pointer) finger, and touch the tip of your nose. Now try the same thing with your left index finger. Keeping your eyes closed, try to touch both fingers together in front of you. Can you do it? Just think of the enormous amount of input flowing into your cerebellum that is necessary for you to be able to do this. It's a good thing most of this information is processed without you having to think too hard about it!

Bossy Brainstem

You probably remember that the brainstem connects the brain and the spinal cord. Like a boss in charge of many people, the brainstem controls a number of functions that are very important, including your body temperature, breathing, heart rate, and blood pressure. You'll notice that these activities are unconscious, or involuntary. You can consciously, or voluntarily, control your breathing to a certain extent – by breathing rapidly on purpose or by holding your breath – but eventually the rate returns to a baseline that your body automatically controls. The brainstem is boss.

The brainstem can produce physical sensations that correspond with your emotions. Has your stomach ever hurt when you were nervous or upset? That's your brainstem's fault. Have you ever seen someone walk by with a refreshing cold water bottle and suddenly felt extremely thirsty? That was your brainstem in action! Do you remember being hungry but feeling even more so when you smelled food? Have you ever wondered why you can stay awake during the day, yet you sleep at night? It's because your brainstem is on the job around the clock!

Sorting Stimuli

The brainstem also acts as a sorting center. Many of the tracts (remember, they are the "nerves" of the brain) passing from the brain to the spinal cord, and vice versa, pass through the brainstem. Like a receptionist in an office, the brainstem connects and directs information from certain areas of the brain to other areas and from certain areas of the spinal cord to multiple areas of the brain.

Another really fascinating thing that happens in the brainstem is a crossover from your left to your right. The signals coming from the cerebrum cross over. This means that the signals originating from the right hemisphere of the cerebrum cross over and enter the left side of the spinal cord. From there they go on to serve the left side of the

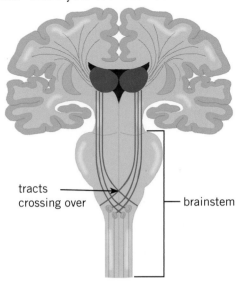

tracts crossing over

brainstem

This is a drawing of a "slice" of the brain and brainstem. The purple lines represent tracts that come from one side of the brain but end up on the opposite side of the brainstem.

body. So the side of your brain that is dominant is the opposite of the side of your body that you write and kick with. Most people are right-handed; therefore, the dominant side of their brain is the left side. Now don't get this confused with whether a person is "right-brained" or "left-brained." If you are right-handed, it just means that the left side of your brain is dominant when it comes to controlling your peripheral nervous system. That has nothing to do with whether you are more comfortable with math (left-brained) or the arts (right-brained).

Try This!

Which of your eyes is dominant? Fix your vision on an object about 20 feet away. Put your hand over one eye. Does your vision shift to one side? Try the other eye. With your dominant eye covered, your vision will shift. When you cover the eye that is not dominant, the object will remain just as you saw it before. The dominant eye chooses how you see the object, and the other eye just enhances what you're seeing by adding some three-dimensional effects. Whichever eye is dominant, the other side of your brain is dominant when it comes to your sense of sight. Therefore, if your left eye is dominant, the right side of your brain is dominant in your sense of sight.

This is just the beginning. In fact, we've only scratched the surface, so to speak. Deeper in the brain, there are a number of other important structures. Hopefully, you will study some of them in high school.

For now, explain all that you have learned so far about the brain.

The Spinal Cord

I've talked a lot about the brain. Now, let's move on down to the spinal cord. The spinal cord is made up of tracts that are safely enclosed inside a long set of bones. Do you know what those bones are? The spinal column! Thirty-one pairs of spinal nerves exit the spinal cord through spaces in the spinal column. Look at the drawing to see how they work their way in and out of the spinal column. As you know, these nerves are responsible for the sending and receiving of a lot of information.

Scientists have named each nerve. You see, since you have 24 vertebrae, each vertebra is given a number. In addition, the sacrum is really five vertebrae fused together, so we can give those vertebrae numbers, too. The nerves are named based on which vertebra they pass by on their way out of the spinal cord. Once the nerve exits the spinal cord, it divides into branches called rami. These rami serve the different parts of the body.

These nerves are like telephone cables that connect your body with your brain. They are made of sensory and motor neurons. The first nerve pair comes from the underside of the brain. All the rest branch out directly off the spinal cord. Each nerve pair controls a particular area of the body. You can probably guess which part of the body each nerve controls by tracing where it goes once it leaves the spinal cord.

Let's explore this a little more. Look at the

The spinal cord is protected by the spinal column. The top vertebra has been removed so you can see the spinal cord.

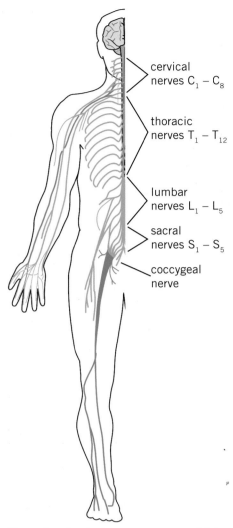

cervical
nerves C$_1$ – C$_8$

thoracic
nerves T$_1$ – T$_{12}$

lumbar
nerves L$_1$ – L$_5$

sacral
nerves S$_1$ – S$_5$

coccygeal
nerve

The pairs of spinal nerves are named for the vertebrae they exit.

drawing on the left. The spine is divided into areas, as you may recall from the skeletal system chapter. The vertebrae in the neck are called cervical vertebrae. The uppermost nerve is named C1, the next one down C2, and so on, through C8. The C stands for cervical. Next come the thoracic vertebrae (remember, thorax means chest). The nerves that come out of these vertebrae are named T1 – T12, with T1 located just below C8. Below the thoracic vertebrae are the lumbar vertebrae. Lumbar refers to the lower back. The five nerves that exit the lumbar vertebrae are labeled L1 – L5. Last of all come the sacral vertebrae. These are named after the sacral bone. The nerves exiting at the sacral vertebrae are called S1 – S5. The last pair of nerves comes out of the coccyx, so it is called the coccygeal (kok' sih jee' uhl) nerve.

Over time, scientists discovered which areas of the body were served by specific nerves. For instance, your thumbs are served by C6. This means that the nerve exiting the spinal column at the 6th cervical vertebra travels down the arm and serves part of the arm, the thumb, and the pointer finger. Sensory information from the thumb travels through a neuron of the same nerve all the way back to C6. From there, other neurons connect to the brain. The sole of the foot is served by S1. This means that the motor and sensory information going to and from the sole of the foot exits and enters the spine at the first sacral vertebra. This has allowed scientists to make a "map" of which spinal nerve serves which area of the body. The map is on the right.

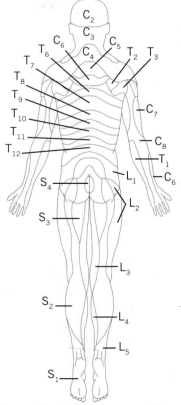

Try This!

Use the map on the right to answer the following questions. If you are having pain in your calf, which nerve is transmitting that message? If you are having pain in your neck, which nerve is affected? Imagine you are having pain in your thumb. Which nerve is telling your brain about that pain? You should have guessed S2, C3, and C6.

Tell someone what you have learned so far.
Be sure to include information about the cerebellum and brainstem!

The red lines are not nerves. They just mark off regions in your body to tell you which nerve serves which region.

The Reflex Arc

You may wonder why the spinal cord is part of the CNS, since the spinal cord doesn't "think" like the brain does. Well, the spinal cord contains interneurons, which actually do process a lot of information. One way the spinal cord processes information is through what we call the **reflex arc**.

Have you ever accidentally pricked your finger with a needle? You probably noticed that by the time you felt the pain, you had already pulled your finger away from the needle. That's the work of a reflex arc. Reflex arcs are made up of a sensory neuron, a motor neuron, and an interneuron. Reflexes are fast, because they don't have to wait for information to travel all the way to the brain. Instead, they use the spinal cord for quick processing.

In the case of pricking your finger with a needle, as soon as the sharp end of the needle hit your finger, the sensory nerves in your finger generated an "ouch" message. That message traveled along the sensory neuron's axon to an interneuron in your spinal cord. The interneuron received the information and activated a motor neuron that sent a signal to some muscle fibers in your finger, causing your finger to pull away. If you had to wait for the pain signal from your finger's sensory neurons to travel all the way to your brain, the needle would have sunk deeper into your skin, causing a lot more damage. So your finger pulled away from the needle even before your brain registered that your finger was being hurt by the needle! There are reflex arcs like this all over your body, ready to sense danger and move muscles quickly!

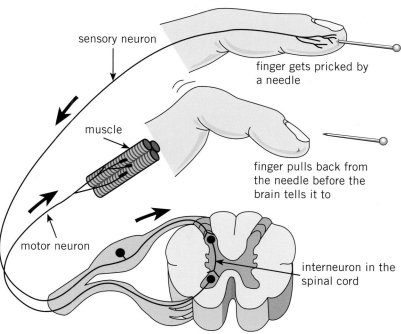

In a reflex arc, an interneuron in the spinal cord decides to activate a motor neuron without waiting on your brain to give instructions.

If you didn't have such great reflex arcs, you would get hurt quite a bit. Think about a time you touched a hot stove. A reflex arc allowed you to pull your hand back before your skin got too badly burned. If you had to wait to realize you were being burned before you pulled your hand back, your skin would have been destroyed! Isn't the design of the human body simply amazing? God worked everything out so we could be safe.

Packaged and Protected

Did you know that your brain is extremely fragile? You might also remember that the brain's neurons cannot repair themselves. In fact, it was once thought that you are born with all the brain neurons you will ever have. While there is now strong evidence that people do add a few neurons to their brains over the course of their lives, the number of added neurons is very small compared to the number of neurons you are born with. That's okay though, because your brain comes with protection. When you think of what protects the brain, you most likely think of the skull. However, the inside of the skull can be rough – rough enough to cut the brain, if it wasn't well packed and padded. Between the skull and the brain is a special fluid called **cerebrospinal** (suh ree' broh spy' nuhl) **fluid**. This fluid helps cushion the brain and keep it comfortably in place.

When you bang your head hard, your brain moves a bit inside your skull. The cerebrospinal fluid's job is to distribute the force of the impact, decreasing the possible movement of your brain within your head. Let's do an experiment to see how this works.

Try This!

For this activity, you will need two real eggs, a plastic Easter egg (that's a little larger than the eggs), and some Karo Syrup (or molasses). Put the first egg inside the plastic Easter egg, and drop it into the sink from about 18 inches above the counter. Open the plastic egg and look at the real egg inside. What happened? Take a minute to clean up the mess. Now, put Karo Syrup in both halves of the plastic egg, and quickly close them around the second egg. You may want to do this over the sink to avoid a huge mess. Now drop the plastic egg from the same height. What happened?

The second egg sustained a lot less damage than the first, didn't it? That's because the Karo Syrup padded the egg. It decreased the egg's ability to move quickly within the plastic at the time of impact, therefore protecting the egg from breaking.

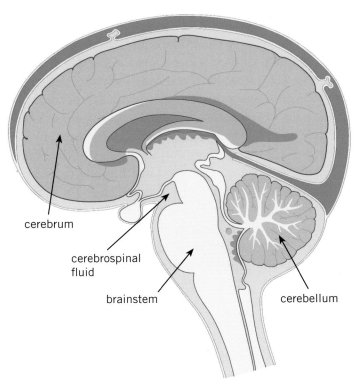

This slice of the skull and neck shows that your brain and spinal cord are surrounded by cerebrospinal fluid.

cerebrum

cerebrospinal fluid

brainstem

cerebellum

Bigger Brains

When you are born, your brain is 75% of the size it will be when you are an adult. Thus, your brain grows as you mature. However, it doesn't grow by adding neurons. Instead, it grows by myelination and by making connections. Myelination is what happens when the fatty tissue myelin forms around the axon of a neuron. It is also what makes your brain grow larger during the first year of your life. It's like adding fat to your head. We all get fat heads as we get bigger. In fact, two-thirds of your brain weight is fat (not just the fat in myelin – there's fat in the cell membranes as well)!

Another thing that contributes to brain growth is the formation of an enormous number of connections, or links, between the neurons in your brain. At birth, there are already many connections between neurons in the brainstem and spinal cord. However, after birth, the number of connections increases quickly and dramatically in your cerebrum. Synapses (remember, those are connections between neurons) seem to be formed when we have experiences. Babies who are not played with or taken places do not form as many connections in their brains as babies that are exposed to a lot of new experiences. The more positive the experience, the more able your brain is to make those connections. However, in childhood the brain is what we call "plastic." This means the young brain can remodel and make up for difficulties by forming connections in other areas of the brain that don't need experiences to grow. Also, by introducing many experiences and learning opportunities in an older child or adult, connections can continue to be made, and the brain can continue to grow in intelligence. God always has an answer for problems we may experience here on earth.

My Brain

You are different from every other individual in the world! But did you realize that the way in which your neurons connect actually contributes to how unique you are? It's true! Your neurons are connected differently from every other person's neurons. Do you remember the dot activity you did at the beginning of this lesson? No one else made the same dot design or the same connections. This is true for your neural connections! That is why you are not exactly like your brothers, sisters, parents, cousins or anyone else in the world. Even if you have an identical twin, you are not really identical to your twin. Your neural connections are just not the same. You are unique. Everyone is born with the potential for a huge number of neural connections. How your brain is fed – both nutritionally, as well as with experiences and information – makes a huge difference as to whether or not neural connections are made, and which connections are made!

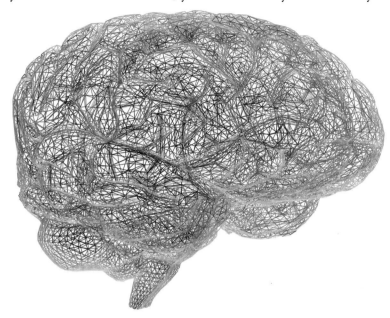

No one in the world has exactly the same neural connections as you do. Because of this, your brain is unique!

The reason you can think and remember information is because these billions of neurons form multiple links with one another. The more links, the more you can store – and more importantly – the more you can recall for later use. Links occur with learning, and not just book learning, but all types of learning. Just think, even as you're learning right now, new links are being formed between your neurons. Then, when you go on and study the human body further, you'll be able to recall this information. As you read new information, your brain will form even more complex links,

allowing you to store and recall even more information! The pathways you use most will form the most links with other pathways. This also means that pathways which are seldom used can fall into poor condition. So keep working on those math facts! You'll need them in high school!

God gave you a wonderful brain. All your thoughts, everything you know about yourself – your ability to draw, dance, create new things, have feelings, learn about God, and worship Him – all take place within your amazing brain. You can exercise your brain by thinking the right thoughts and making the right choices. In fact, you can train your brain to think correctly by doing this.

The Bible tells us, "*And do not be conformed to this world, but be transformed by the renewing of your mind, so that you may prove what the will of God is, that which is good and acceptable and perfect*" (Romans 12:2). This command is especially interesting in light of all you have learned in this lesson. You see, as we learn more about God's ways and practice thinking His thoughts, we are laying down new neural pathways in our minds – new ways of thinking – new neural connections! We are truly transforming our lives because we are creating new ways of thinking. As Colossians 3:2 instructs, "*Set your mind on the things above, not on the things that are on earth.*"

If you continue to learn about God and pray to Him, you will form new neural pathways that will lead to transformed thinking and a transformed life!

What Do You Remember?

Name the four lobes in the cerebrum. Which lobe is responsible for speech and language? Which lobe is responsible for your emotions? Which lobe processes smell and memory as well as tone and loudness? Which lobe processes sensory information and integrates it to determine where you are in relation to your surroundings? Which lobe processes visual information? Which side of your brain is more active when you are doing math? Which is more active when you are working on a piece of art? How does myelin help your neurons? Which part of the brain is responsible for keeping you balanced? What is the reflex arc? Why are interneurons and interconnections between neurons important?

Notebooking Activity

Record all that you've learned about the brain and spinal cord in your notebook. Try to include information about the different lobes of the cerebrum, the two hemispheres in the brain and what they control, how myelin helps your nerves, what your cerebellum does, and how your brainstem works. Also include information about your spinal cord and the different nerves you have exiting it, the reflex arc, and how God designed protection for your brain, as well as how your brain grows as connections are made. By jotting down this information, you'll add lots of neural pathways to your brain!

Project
Design a Science Fair Project

It's always a good idea to practice designing your own science experiment. Today, you will have an opportunity to do this. After you design your experiment, you might consider entering it into a science fair. There are a lot of rules that govern science fair projects, so you will need to find out what those rules are and make sure you follow them exactly. For example, if you use people, children, or animals in your project, you will have to follow specific guidelines given by the science fair.

Your experiment should use the scientific method with controls and variables. A control is something that you will compare your results to. For example, if you want to test whether or not a certain food makes a person smarter, you have to do something that measures the person's intelligence before he starts eating the food. That is your control, because once he starts eating the food, you will compare your measurements of his intelligence to that measurement. If the person's intelligence measurements increase compared to the measurement before he started eating the food, then the food might have made him smarter. A variable is the part of the experiment that changes. In the experiment I just described for example, you changed the food the person was eating. That's the variable. For an experiment to be really good, there should be only one variable. Everything else should stay the same.

You will want your science experiment to explore some aspect of the brain. On the next two pages, you will find some ideas to get you thinking:

Experiment Ideas

1. You could design an experiment that measures whether boys or girls have better hand-to-eye coordination.

   a. Have the boys and girls race remote control cars in a large maze. Measure the time it takes to complete the course, the only variable being the child doing the race.

b. Have the boys and girls throw a ball into a small square hole in a box several feet away. Record how many tries it takes each child to get the ball in the hole.

For either experiment, you would compare the average result for the boys to the average result for the girls. If the average result for the girls was better (a faster time through the maze or fewer tries to get the ball in the hole), it might mean the girls have better hand-to-eye coordination than the boys.

2. You could design an experiment to measure whether adults or children learn new skills more quickly.

c. Have the adults learn a series of hand motions. Measure how long it takes for them to learn the entire series. Then have children learn the same hand motions and see how long it takes them.

d. Have the adults build a card house and see how long it takes for them to get the job done. Then have the children build the same-sized card house and see how long it takes them.

For either experiment, you would compare the times for the adults and children. If one group's average time was shorter, perhaps that means they learn faster.

Measuring how long it takes to learn a series of hand motions is one way to measure how quickly a person learns.

3. You could design an experiment that measures whether boys or girls have better short-term memories. A person's short-term memory refers to the things he or she remembers shortly after being exposed to information.

e. Place various items on a tray and allow the boys and girls to look at the tray for 30 seconds. Then hide the tray and have them recall what they saw.

f. Recite a list of items, and then ask the boys and girls to repeat them back.

Seeing how well people can remember a bunch of unrelated items on a tray is a good measure of short term memory.

For either experiment, you would compare the number of correct answers each group gave. If one group had a better score, that could be because the group was made up of people with better short-term memories. A variation could be to compare age groups rather than genders. If you can find a few children aged 5-7, for example, and others whose ages range from 10-12, you can see if the older or younger children have the better short-term memory.

4. Design an experiment that utilizes food and its effects on learning, memory or motor skills.

g. Do any of the experiments above, but make the variable a change in the diet.

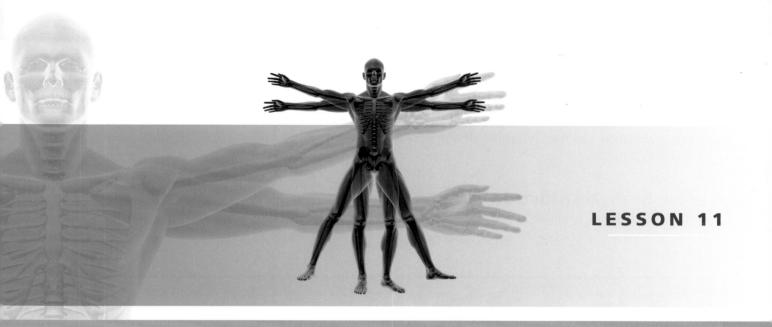

YOUR SENSES

You've learned a lot about how your body senses and processes information. Yet, can you name all of your senses? You may have learned about them already. If you didn't, that's okay. You can probably figure out most of your senses if you think about the main ways in which you experience the world around you. Imagine walking into your house after a long day of playing outside. You haven't eaten for hours. Your mom is cooking dinner and tells you to wash up. What are some of your senses that are active? Well, you hear your mother's happy voice (unless you're late, then her voice may not sound so happy). You smell food cooking. You see the kitchen table set for dinner, and you feel the warm soapy water cleaning the dirt off your hands. Finally, after you've said grace, you'll taste your delicious meal! Ahhh! Aren't your senses simply divine?

Traditionally, students are taught that the five senses are **sight, hearing, smell, taste**, and **touch**. However, that's not the most scientifically accurate way to classify your senses. The best way is to say that you have one general sense, which is your sense of touch. We call it a general sense because it occurs all over your body. Then, you have five **special senses**, which are smell, sight, hearing, taste, and **balance**. We call them special senses because they occur as a result of specific organs at special places in your body. In other words, you can't experience the sense of sight over your whole body. You can only see from your eyes. That makes it a special sense – one dependent on your eyes. In contrast, your whole body can experience the sense of touch, so touch is a general sense.

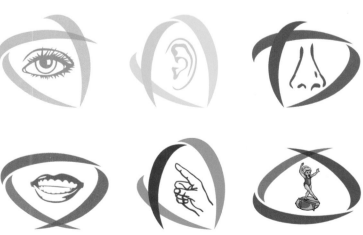

The five color drawings represent the five traditional senses. A more proper classification, however, calls the sense of touch a general sense and the other senses (including balance) your five special senses.

However you classify them, you can give credit to the afferent neurons in your peripheral nervous system for your senses. They deliver all this wonderful input to your central nervous system and enable you to actually *experience* the world around you in different ways! In this lesson, we will explore the five special senses: smell, taste, balance, hearing, and sight. In the next lesson, when we discuss skin, we'll spend some time on the general sense of touch.

Old Fashioned Olfaction

I'm sure you are aware of the fact that your nose is the location for your sense of smell. But did you know that your nose also has a major impact on your sense of taste as well as on your voice?

Try This! Take a bite of something you like to eat. Enjoy it in your mouth and swallow it. Now plug your nose and take another bite. It doesn't taste as good, or perhaps I should say, it doesn't taste as much, does it? Without your sense of smell, the taste of food is less powerful. It's still there, but it's rather bland. Now plug your nose again and speak. Wow. Your voice doesn't sound quite as lovely, does it?

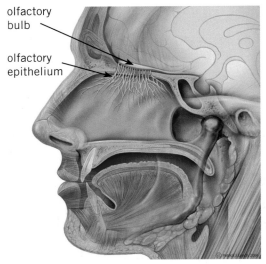

olfactory bulb

olfactory epithelium

The epithelium has millions of tiny hairs that float in the mucus layer of your nose picking up the odors flowing into your nostrils.

Your nasal passages are quite important, aren't they? Believe it or not, your nose even affects your sleeping positions. If you're lying on one side, the lower nostril will slowly fill with mucus. This causes you to roll over so that the nasal cilia will be able to fight gravity and move the mucus to the back of your throat to be swallowed.

Your sense of smell is accomplished with your **olfactory** (ol fak' tuh ree) **system**. That's a fancy word that means the system that gives you your sense of smell. In the roof of the upper nose is the **olfactory epithelium** (ep' uh thee' lee uhm). This is a cluster of smell sensors that contains about 10 million olfactory cells. Since you have two nostrils, you have 20 million smell cells! That may seem like a lot. It sure does to me. Yet dogs have over 100 million smell cells. That's a lot more. No wonder they are used to locate lost people, bombs, and illegal drugs!

Surface area, surface area. You know how much God likes surface area! Well, it turns out that each olfactory cell has several tiny hairs (**cilia**) projecting from its tip, greatly increasing the surface area so that even faint odors can be detected. These tiny hairs float in the mucus layer produced by **olfactory glands**. The cilia sense odors coming in, even if you have a lot of mucus. Of course, if you have too much mucus, they get covered up and hinder your sense of smell. Have you ever had a bad cold and couldn't smell or taste your food? Your cilia were drowning in mucus and couldn't sense odors.

Once an odor reaches the cilia, the olfactory cells send

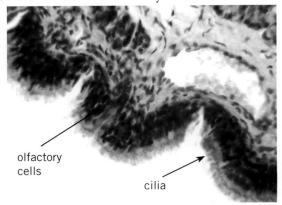

olfactory cells

cilia

This magnified image shows you the cilia and olfactory cells of an olfactory epithelium. If no molecules reach these cilia, or if the cilia do not sense them, you do not experience smell.

signals to neurons in your olfactory bulbs, which take the information to your brain. How do the nerves get through the bone? Well, God designed the cranium to have tiny holes in it. The nerves pass through these tiny holes so they can take signals to their proper destinations. The olfactory bulbs carry the information directly to the temporal lobe of the brain. You remember that the temporal lobe processes information from the sense of smell, don't you?

So what starts this process? Well, odors (things you smell) are actually chemicals that are floating in the air. These chemicals are typically gases. They enter your nose and dissolve in the mucus of your olfactory epithelium. This activates the cilia of the olfactory cells, causing them to send a signal that ends up in your brain. No one has yet figured out the details of how the brain can tell one smell from another – how it can identify the smell of a rose and distinguish it from the smell of garlic. Yet most people can identify about four thousand different smells, and those with particularly gifted noses can smell up to ten thousand smells. Those with this gift often end up working as wine, perfume, or food testers. They are also the ones who may complain that your cologne is too strong! Believe it or not, there are even special training schools for those who can smell well. They help these people refine their olfactory skills. They are taught to "sniff" smells, as well as inhale them naturally. This is because when you sniff, air enters your nose far more rapidly than when you breathe normally. The incoming air travels more directly upward, with more of the air coming into contact with the olfactory epithelium before it diffuses throughout the nose.

Try This!

You can try this yourself! Get out a bottle of vanilla. Open it, and hold it a few inches from your nose. See if you can pick up the scent while inhaling naturally. Using your hand, wave some of the air over the opened bottle towards your nose, still inhaling naturally. Now, hold the bottle directly under your nose and sniff. Can you smell the difference?

You've probably noticed that not everything has a smell. This can be because something doesn't emit gas molecules, so nothing reaches your olfactory epithelium. Your kitchen utensils don't have a smell (when they are clean), because they don't emit molecules, so there is nothing from them to inhale. However, a substance can also be odorless if your olfactory cilia cannot sense the molecules it emits. Natural gas has no odor at all, even though it is easy to inhale. That's because your olfactory cilia can't sense it. Because it is a very dangerous gas, the gas companies add a very fragrant chemical to natural gas in case there's a leak. This smell is very recognizable and helps keep people safe. If natural gas is leaking, it can be fatal.

Your sense of smell can diminish after a while. I'm sure you have noticed this yourself. When you first walk into a room that contains a strong smell, it's about all that you notice. But after a while, the smell will often become less obvious. If you leave the room for a while and then return, the odor will be quite strong again (unless someone has opened the windows!). This is because your olfactory cells will stop sending signals to your brain after smelling the same thing for a long time. The smell hasn't gone away; your olfactory cells just stopped responding to it.

Try This!

Because your sense of smell is connected with your memories in your temporal lobe, it should be easy to learn and remember how items smell. Let's test this out using the herbs in your kitchen. Choose several herbs from your parents' cabinet. Ask them to help choose ones that have distinct odors. Place four different herbs on separate paper plates, and write the name of the herb in large letters on the plate. Spend some time sniffing and smelling each. Then, close your eyes and have a family member mix up the plates or bring you one at a time. Now sniff again. Can you identify each herb?

Have you ever wondered if God can smell things? Well, Deuteronomy 4:28 says that idols (false gods) cannot "*see nor hear nor eat nor smell.*" The point of the passage is to indicate how worthless idols are compared to the real God, so that indicates God can, indeed, smell. Also, Ephesians 5:2 says that Christ's sacrifice for us on the cross was like a "*fragrant aroma*" to God.

Revelation 5:8 says that God has "*golden bowls full of incense, which are the prayers of the saints.*" Do you know what incense is? It's a substance that has a strong odor when it is lit. There are lots of different kinds of incense, and many kinds smell quite lovely. So the Bible implies that God collects our prayers, and they smell lovely to Him. Imagine that! He loves it when you offer these sweet smelling prayers to Him! Be sure to spend some time in prayer today.

The Bible compares our prayers to incense, which produces a sweet-smelling smoke. You can think of your prayers as rising up to God as a sweet-smelling offering to Him.

Explain in your own words what you know about the sense of smell.

Tasty Taste Buds

"*O taste and see that the LORD is good; How blessed is the man who takes refuge in Him!*" (Psalm 34:8). This psalm uses poetry to describe how wonderful God is to His people. Taste is such an important sense! It is a wonderful gift from God, because it brings so much pleasure to our lives. Have you ever savored a piece of food? Maybe you've had the experience of slowly eating a warm chocolate-chip cookie, intensely enjoying the taste of every bite! Think of how wonderful it is to taste your favorite dessert. God could have created humans without the ability to taste such delightful things. He could have created us so that we only eat to satisfy the grumblings in our stomachs. But instead, God desired to give us joy in eating by designing taste as part of our eating experience. He did this because He loves us and wants to fill our lives with abundant joy! He is good. Taste and see!

What part of your mouth do you think does all the tasting? It's your tongue, of course! Your tongue has a lot of jobs. You might remember that your tongue is important for talking. Do you remember learning from the digestion lesson that your tongue is extremely valuable for moving the food around your mouth to form the bolus that you will swallow?

Try This!

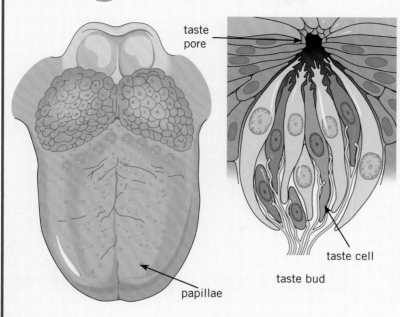

taste pore

taste cell

taste bud

papillae

Certain bumps on your tongue have tiny holes in them called taste pores. They lead to the taste buds, which allow you to taste food.

Get a mirror. Now watch your face and mouth as you recite the alphabet. But instead of saying the names of the letters, say the sounds they make. Instead of saying "Ay," say "ah" for the short sound the letter "A" makes. Instead of saying "Bee," say "buh" for the sound the letter "B" makes. You'll notice, in addition to your tongue, that your lips, facial muscles, and jaw are all necessary for forming sounds.

Look at the bumps on your tongue. They are called **papillae** (puh pill' ee), and some of them house your taste buds. You have about 10,000 taste buds. They detect five specific taste sensations: **sweet, sour, salty, bitter, and umami** (oo mah' me). Some call that last one "savory;" it is the taste of meat. All tastes you experience are a combination of those five taste sensations. For many years, scientists believed the taste buds were located in specific areas of the tongue. They thought a person could taste sweet tastes in one area, bitter in a different area, and so on. But today, scientists don't think that's true.

Try This!

Let's investigate the old idea that the taste buds are located in specific places on the tongue. You will need five small custard cups, five Q-tips, salt water (salty), a lemon (sour), some sugar (sweet), and unsweetened cocoa or ground coffee (bitter) that has been moistened with water. First, place each substance in each of the five cups. Rinse your mouth a few times with water before starting. You will also want to rinse in between each test, in order to keep your taste buds clear. Now, look at the diagram showing where scientists used to think each taste bud section is located. Dip a Q-tip in one of the cups, and use the substance to test the various areas of your tongue. Using a new Q-tip for each cup, test the other substances. If the old idea was correct, you shouldn't taste salt water very well in the areas that are supposed to have sweet, sour, or bitter taste buds. However, you should taste it strongly in the area that was thought to contain salty taste buds. Is this what your results indicate? If not, then scientists were right to abandon this idea.

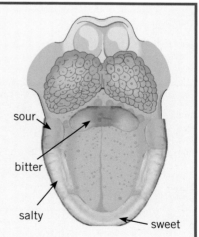

sour

bitter

salty

sweet

Scientists used to think that taste buds were concentrated in certain regions.

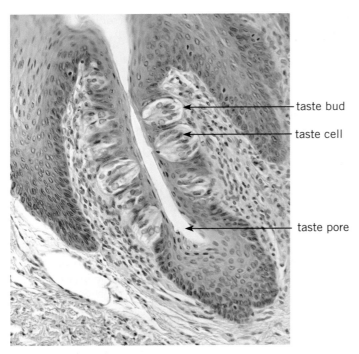

taste bud

taste cell

taste pore

This microscopic photo shows you taste buds in the tongue. Chemicals must reach them to activate the sense of taste.

Dissolving Donuts

In order to actually taste food, the chemicals in the food must dissolve in the saliva and pass into the taste pores. As with the olfactory cells, the taste (also called gustatory) cells require a moist environment. They also end in tiny hairs that increase the area available for receiving taste. These hairs send a signal to the taste neurons, which then send the signal on to the brain.

Can you think of a reason why God created your tongue with a bunch of bumps (papillae) instead of smooth, like your fingers? Well, as is the case with the brain and intestines, this design increases the surface area greatly. It gives us a keen sense of taste packed into a smaller area. Have you ever felt a cat's tongue? It is very rough, isn't it? That's because it has a lot of papillae on it. This should tell you that taste is a very important sense in cats. Perhaps that's one reason cats have a reputation for being finicky eaters.

Get It While It's Hot

The reaction of your taste buds depends on temperature. In general, warmer food stimulates your taste buds more than cold food. As a result, tastes are usually more intense when the food is warm. Now that may not mean you enjoy the food more, but it usually has more taste. This is because heat causes molecules to move faster. This increases the amount of chemicals that dissolve in your saliva, and it also increases the speed of the chemical reactions that operate your taste buds. When something is cold, its molecules move more slowly, which means they don't dissolve as well in saliva. In addition, the chemical reactions that operate your taste buds don't run as quickly. Let's find out if you can experience this effect.

Try This!

Get a few saltine-type crackers, a glass, a mug, and some chocolate milk. Warm up some of the chocolate milk in the mug. Now take a sip of the warm chocolate milk. Consider the taste. Then, eat a saltine cracker to remove the taste from your mouth. Next, take a sip of the cold chocolate milk. Which one had more taste to it? You can try the warm milk again to help you compare.

You already know that your sense of taste is connected to your sense of smell. Your smell receptors are activated when you bring food near your face. This is how taste and smell interact. Have you ever had a bad cold and found it difficult to taste your food? That's because you are used to smelling the food, which enhances the taste.

Sometimes when people have had head injuries, the tiny olfactory nerves leading to their heads are torn. When this happens, these people lose their sense of smell, and along with that, much of their sense of taste. In addition, damage to certain regions of the temporal lobe can cause a loss or reduction in the senses of smell and taste. Also as people get older, the olfactory cells degenerate, which can cause a loss in the sense of smell. As a result, the sense of taste is not as strong, either. This is why some older people sometimes use more salt on their food and more perfume on their bodies. This is very sad, and it often causes a loss of interest in eating, since eating loses much of its God-given pleasure when you cannot smell your food.

Try This!

If you want to explore just how important smell is for taste, try the following experiment. Gather some foods with familiar tastes and similar textures. For instance, try an apple, a potato, and a radish. Cut the foods into similarly bite-sized pieces. Now, close your eyes and plug your nose tightly. Have a friend or sibling feed you a bit of each of the foods and see if you can identify them. Try it again without your nose plugged. What do you find? If you'd like, you can repeat this experiment using apple juice, fruit punch, and cranberry juice.

Explain what you have discovered about your sense of taste.

Now Hear This

You know that the ear is the sensory organ for hearing, but did you know that the ear is also seriously important to your sense of balance? Both the sensory organ for hearing and the sensory organ for balance are located in the inner ear. Let's explore the structure of the ear and then discuss hearing and balance. The ear is divided into three parts: the **external ear**, the **middle ear,** and the **inner ear.** As I describe the structure of the ear, look at the diagram to see where each part is located.

External Ear

The **pinna** (pin' uh), also called the **auricle** (or' ih kuhl), is the part of the ear you see from the outside. It is like a funnel, guiding sounds into the **external auditory** (ah dih' tor ee) **canal**. This is the part of your ear that gets wax in it. You see, instead of being lined with mucous glands like the nose, the external auditory canal is lined with wax glands. The wax works a lot like mucus. When something gets in your ear, wax glands secrete extra wax to trap it. This way, dust and germs don't make their way into your middle ear. The external auditory canal leads directly into the skull, but it "dead ends" at the **eardrum**. The pinna, external auditory canal, and eardrum make up the external ear.

External Ear
pinna, external auditory canal, eardrum

The ear is responsible for both the sense of hearing and the sense of balance.

Middle Ear

Just on the other side of the eardrum is the middle ear, which is filled with air (unless you have an ear infection). Inside your middle ear are the tiny bones mentioned in the skeletal system lesson. These bones are the malleus, incus and stapes. We also call these bones the hammer, anvil and stirrup because each bone resembles those items. The malleus resembles a hammer, the incus resembles an anvil and the stapes resembles a stirrup. You can look up these items on the Internet and see if you agree!

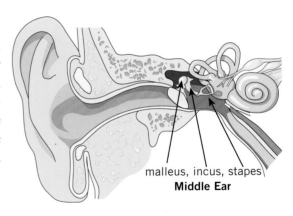
malleus, incus, stapes
Middle Ear

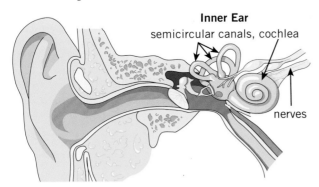
Inner Ear
semicircular canals, cochlea
nerves

Inner Ear

Let's go a little deeper into your skull. There you'll find your inner ear. Inside the inner ear, you will find the **cochlea** (kok' lee uh) and three **semicircular canals**. Look at these in the diagram on the left. All three of these structures contain fluid. The cochlea allows you to hear, and the semicircular canals help to control your balance.

Hearing in a Nutshell

Did you know that sound travels in waves? These waves are kind of like the waves that span out from a pebble dropped in a pond. The sound waves actually vibrate particles in the air, causing "clumps" of air to form. As long as the sound waves have something to travel through (like air), they will continue to move along. If your ear is in the way, the waves will travel into your ear.

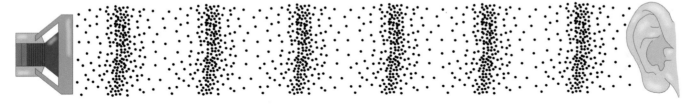

The sound from the speaker vibrates the air, forming clumps of air molecules (represented by the black dots). When those clumps reach your ears, you will hear the sound.

Try This!

To get an idea of how sound waves travel, get out a Slinky. Stretch it out along a flat surface (like a tabletop), holding each end in one of your hands. Keep your right hand still, and start moving your left hand back and forth, first towards your right hand and then away from it. Keep "pumping" the Slinky with your left hand and look at what the Slinky does. It clumps together in regions, and those clumps move from your left hand to your right hand. That is how sound waves travel in air. By vibrating the air, sound forms "clumps" that travel away from where the sound was made.

This should tell you something. In order for sound to travel, it must have something to vibrate. It can vibrate air, water, or a solid – anything that has molecules in it. However, without something to vibrate, sound cannot be made. For example, in space there is nothing to vibrate. As a result, it is totally silent in space. Even if you talk, your voice will not travel out of your space suit and into space, because there is nothing for the sound to vibrate!

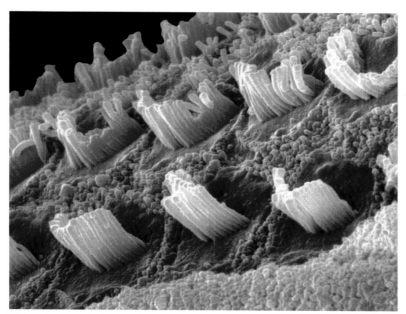

This microscopic image (color artificially added) shows the tiny "hairs" that project from cells that line the cochlea. Vibrations in the fluid of the cochlea cause the "hairs" to move, which causes a signal that can be sent to the brain and interpreted as sound.

Now that you know about how sound travels, let me explain to you how hearing works. When sound waves hit your pinna, they are funneled into the external auditory canal, where they eventually bounce against the eardrum, causing it to vibrate. This vibration is passed along to the auditory ossicles. God placed the bones just so, and this special placement causes them to amplify the vibrations about 20 times!

Once the vibrations are amplified by your ossicles, the stapes passes the sound waves to the cochlea. You may be surprised to learn that sound travels well through water. In fact, whales can speak to one another from thousands of miles away because of this. Well, God knew this and filled the cochlea with fluid. So the vibrations pass on through the fluid in the cochlea, causing the molecules of the fluid to vibrate. The vibrations travel through the cochlea, and eventually come in contact with the "hairs," or tiny hair cells that line the cochlea. The "hairs" move in response to the vibrations, and that movement is converted into a signal by the hair cells. The signal is then passed on to the cochlear nerve, which goes to the brain. The brain receives the signal and then interprets it, so that you end up hearing the sound.

Sound Off

Having two ears allows you to pinpoint the location of sounds. The waves from a given sound travel at the same speed, but your ears are in slightly different locations. Your brain can process the information from one ear, compare its time of arrival to that of the arrival at the other ear, and give you an excellent idea of where the sound is coming from! The sound is also louder in the ear closer to the sound, and your brain uses that information as well.

Try This!

To demonstrate this, have a friend sit in the center of the room blindfolded. Tap a pencil on an empty cup from different areas of the room. Have your friend point to where the sounds are coming from. Now try it again, but this time in addition to the blindfold, have your friend cover one ear. How accurate is your friend in identifying sounds when one ear is covered?

If you listen to loud music or other loud noises over and over again, the hairs of your hair cells can be damaged. Even a single very loud noise can damage them. When the hairs are damaged, you lose some (or all) of your ability to hear. You may lose your hearing only temporarily if the damage is mild, but there is also a risk of losing your hearing permanently. A friend of ours was in a rock band for many years. Today, he can only hear very loud voices. He cannot hear children talking. This is why you should not listen to music at full volume. Protect your hearing! It's a great gift from God.

Anything that keeps the eardrum and ossicles from moving sound along will lead to a decrease in hearing. An ear infection fills the middle ear with thick pus. This results in decreased hearing, because the vibrations cannot be amplified. When my son was a baby, we learned that he could not hear because of chronic ear infections. So, he had to have a small plastic tube placed in his eardrum to drain out the fluid. The tube eventually fell out; his eardrum healed from the hole placed in it, and his hearing returned to normal.

Sound can travel through anything that has molecules to vibrate, including bone. This is why you can hear yourself chewing. Unless you are chewing with your mouth wide open, no one else hears your chomping as well as you do. Have you ever noticed that when you hear yourself talk on a recording, your voice sounds different? That's because a large part of what you hear from your own voice is the result of your sound waves traveling through your bones to get to your ear. Since they are traveling through bone, they are a bit different from when they travel through air. As a result, you sound different to yourself than you do to others.

All Fall Down

Stand up and try to stand on one leg. How long can you do it? With practice it becomes easier and easier. Baseball pitchers and ballerinas have to stand on one leg quite a bit. Their sense of balance makes this possible. There are two kinds of balance, static and dynamic. The static sense tells you the position of your head, whether it is tilted up, down, or sideways, or is looking straight ahead. The dynamic sense informs you about active movements of your head, like going around a corner on a bike.

Let's start with the static sense, which is located in the **vestibule** (ves' tuh byool) of each inner ear. The vestibules each contain two sacs. Within each sac is a small spot called the **macula** (mak' yuh luh), which is the Latin word for spot.

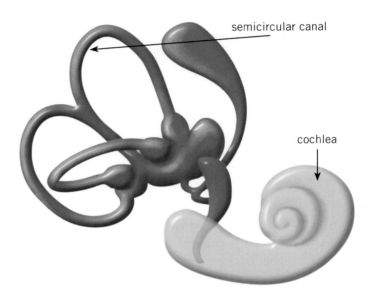

semicircular canal

cochlea

Your sense of balance is housed in your inner ear (colored in purple).

Each macula is made of a thick, gel-like fluid, and inside the fluid are teeny-tiny stones. Yes! Stones! God placed stones in your ears. They are called **otoliths** (oh' tuh liths) – "oto" for ear, and "lith" for stone. Ear stones! Why on earth did God give us ear stones? Well, stones are heavier than fluid, and when you tilt your head, gravity pulls on the ear stones. There are tiny nerve cells, called hair cells, embedded in the macula, and when gravity pulls on the ear stones, the macula moves. That bends the hair cells, which causes them to send a message to the brain. The brain then knows which way your head is tilted.

The sense of dynamic motion is detected by the semicircular canals. Each of your semicircular canals has fluid in it that can flow around in a circle if you get it moving. One part of each semicircular canal is widened, and within that space is a tiny sail-like structure called the **cupula** (kup' yoo luh). The cupula is made of gel, but there are no stones in it. Like the macula, though, it has tiny hair cells embedded in it. When the head moves actively, the fluid in the semicircular canals starts flowing, and the cupulas are pushed by the fluid. The very sensitive hair cells in the cupulas are bent by the movement, and they send a message to the brain. Several parts of the brain receive information from your semicircular canals as well as from your vestibules. The brain also receives information from your eyes, from the soles of your feet, and from things that you touch, like a handrail on a stairway. All of this information helps the brainstem and the cerebellum to keep you balanced and to fine-tune your movements.

What happens when you spin in circles for a while and then stop? You get dizzy, don't you? Well, even after you stop moving, the fluid in your semicircular canals keeps flowing, causing your cupulas to move, which stimulates your hair cells. That's why you sometimes lurch or fall down when you stop spinning. The input from your semicircular canals falsely tells your brain you are still moving! The room seems to spin around,

too, because the brain sends messages to keep your eyes moving, as if you are really still moving!

Ballet dancers use a method called spotting when they turn. This means they fix their eyes on one spot, which they try to keep in focus throughout the turns. When they must turn their heads away from the spot, they whip their heads around and find the spot again, keeping it in focus through the turns. They continue this whipping of the head to keep the spot in focus. This enables them to remain balanced while turning. It shows that the eyes are very important to the sense of balance. You can try this yourself.

Spotting while you turn around and around keeps you from getting overly dizzy.

Try This!

You will want to use a timer to see how long it takes for you to feel normal again after you stop turning. First try turning without spotting. Turn in place seven times while keeping your head still. As soon as you finish turning, start the timer and stop it when you feel normal again. Now, try the same procedure while spotting. Find a spot slightly above your head to fix your eyes. It's helpful if it is actually something you can focus on, like a picture. Now, start turning, but try to keep that spot always in view. Whip your head around to get it back in view when you turn your back to it. This takes practice, so you may have to try it a few times to keep the spot in focus. Once you've got it, turn in place seven times (don't forget to spot!), stop, then start the timer. Stop the timer when you feel normal again.

What happened? You were probably still dizzy, but you may have noticed it took less time to get back to normal than after the first set of turns. Go to the course website I told you about in the introduction to the book to see some videos of ballet dancers spotting. You will be amazed at how many turns they make. What's even more amazing is these dancers continue right on dancing!

Tell someone all that you've learned about hearing and balance.

Seeing is Believing

Now we're going to look at the sense of vision. You've learned how amazing your senses of taste, smell, balance, and hearing are. Well, your sense of sight is even more impressive! It's so amazing that it truly gives evidence for the genius of God's design. As Dr. David Menton, professor emeritus at the Washington University School of Medicine says, "The Bible tells us that God's eternal power and divine nature are clearly seen in the things that He has made. One of the most obvious displays of His creative power is the human eye." [http://www.answersingenesis.org/articles/am/v3/n3/seeing-eye, retrieved 08/13/2009]

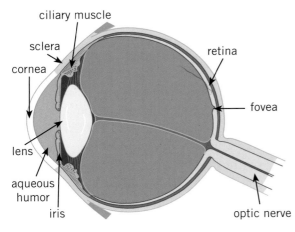

ciliary muscle

sclera

cornea

retina

lens

fovea

aqueous
humor

iris

optic nerve

The human eye is a marvel of design engineering!

Eyeball to Eyeball

Your eye is a ball. Indeed, that's why we call it an eyeball! If you could remove your eye from your eye socket, it would look like a ball covered with a layer of fat that cushions and protects it. Let's look at an illustration of the eye to identify some of its parts.

The white outer part of your eyeball is called the **sclera** (sklir' uh). It's actually a tough, thin bag that forms your eyeball. The **cornea** (kor' nee uh) is the clear "window" in the front of your eye. It is attached to the sclera. Light can pass through the cornea and into your eye. Together, the cornea and the sclera form a tough protective covering for your eyeball. Close your eyes and gently feel one of your eyelids. Do you feel a bulge in your eyeball? That's your cornea. It bulges out a bit. Move your eyes from side to side with your eyelids closed. Can you feel the corneas move?

The blue, brown, or hazel-colored circle behind your cornea is your iris. The round black spot in the middle of your iris is your **pupil**. Light enters the pupil. The iris automatically adjusts the size of the pupil, making the pupil bigger if you are in dim light and smaller if too much light is entering the eye. Do you think the muscles of the irises are smooth or skeletal? Here's a hint: Are they under voluntary control or involuntary control? Do you know what your irises are doing the way you know that your hand is turning the page of this book? That should tell you they are made of smooth muscle.

A clear, watery fluid called **aqueous** (aye' kwee us) **humor** moistens the back of the cornea and the iris. Aqueous humor flows right through the pupil and moistens a clear, oval structure behind the pupil, the lens of your eye. Like the lenses of eyeglasses, your eyes' lenses help you to see near and far. Smooth muscles adjust the lenses so that you can focus on the objects you want to see. Unlike the rigid lenses in people's glasses, however, the lens in your eye is very flexible (at least until a person reaches age 45 or so). When your focusing muscles are contracted, the lens becomes rounder and fatter, which allows you to focus on nearby objects. When they are relaxed, the lens becomes flatter, which allows you to focus on distant objects. This is why looking at scenery off in the distance is more relaxing than looking at something up close. When you focus on distant objects, your focusing muscles are relaxed. It also explains why people need reading glasses as they age. The lens becomes less flexible and remains flat. That makes it impossible for the person to focus on small, near objects.

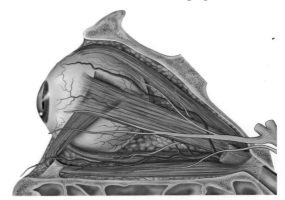

Notice that there are muscles attached to the eyeball. This enables you to look around and roll your eyes. What else do you notice about this eye?

Try This!

To see your irises in action, take a flashlight into a darkened room with a mirror (a dark bathroom works well). Give your eyes a minute to adjust, then using as little light as possible from the flashlight, look at your pupils in the mirror. Now turn on the lights and check your pupils again. What do you see?

Lining the inside of your eye is your **retina**, which has special cells called rods and cones. These cells allow you to see in low light and in color. These amazing cells convert light into electrical signals, sending them along a nerve called the optic nerve to the brain. The cones enable you to see in color; the rods enable you to see in dim light but don't give you color vision. When light is low, then, you see only in shades of gray. You have about 7 million cones and up to 150 million rods. Nocturnal animals have more rods than humans do, so they can see well at night. It's estimated that owls have 100 times more rods than we do and can see as well at night as we can see during the day! They don't have as many cones, however, so they can't see color like we do.

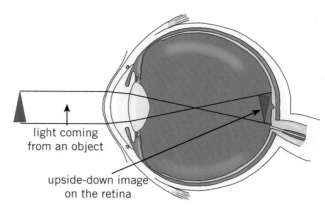

light coming from an object

upside-down image on the retina

Notice that because of the process of focusing light on the retina, the image formed there is upside down. We will talk about that shortly.

The rods are spread all over your retina, while the cones are more concentrated in one spot, called the **fovea** (foh' vee uh). When light from an object enters the pupil, it passes through the lens and is focused on the retina. Whatever is at the center of your vision gets focused onto the fovea. The cones there respond to the light that is being focused on them and send signals down the optic nerve and to the brain. The brain then interprets those signals and forms an image, which is what you see.

Try This!

Let's test your night vision and your color vision. You will need colored pencils (a box of crayons would also work) and one piece of paper for each individual doing the test. Take your pencils and papers into a room that is poorly lit. Can you identify the different colors? On your piece of paper, write down the color you think you have in your hand. If you believe you are holding a red pencil, write down red. If you believe it is brown, write down brown. After every person has tested the pencils, turn on the light and see who had the most accurate vision.

Color My World

Some people are born **colorblind**. That doesn't mean they can only see in black and white. It means that they do not see all the colors that they should. Depending on how colorblind the person is, he might be able to see only a few shades of green, or he might not be able to tell the difference between green and red! Can you think of a few things that would be difficult to do if you were colorblind? Can you think of a few things that you don't need color vision to do? People who are colorblind learn to cope with their vision in many different ways. For example, when driving, they have to remember where the red and green lights are located on the traffic light.

Cornea Control

Although the lens in your eye takes care of some of the focusing of light onto the fovea, the cornea is actually in control of more than half of your focusing power. The cornea does not change shape like the lens, but because it is curved, it automatically bends the light that comes into your eye. The lens then adjusts to correct the focus for different distances. If your cornea is not shaped correctly, you'll have a visual defect called **astigmatism** (uh stig' muh tiz' uhm). With astigmatism, the light entering the eye cannot be brought into sharp focus because it is not all bent correctly by the cornea. As a result, everything is a bit blurry. Special glasses can correct astigmatism by correcting the way the light is bent before it hits the eye's lens.

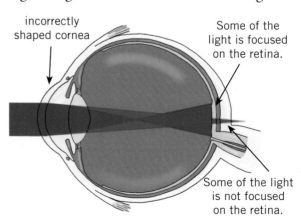

incorrectly shaped cornea

Some of the light is focused on the retina.

Some of the light is not focused on the retina.

In astigmatism, the cornea causes light that enters the eye at some angles to be bent incorrectly. As a result, not all of the light can be focused onto the retina, and the image is blurry.

Try This!

To see how your cornea and lens work together to focus light onto your retina, get a magnifying glass, a sheet of paper, and a pencil. Put a small dot on the sheet of paper. Then go outside and try to catch the light from the sun with the magnifying glass in such a way as to focus it on the dot you made. Do you see how the light narrows down to a point? That's what your cornea does – it automatically bends the light so it narrows to a point. You needed to adjust the magnifying glass back and forth, this way and that way, to get the light focused right on the dot, didn't you? This is what your lens does for you! It doesn't move back and forth. Instead, it changes shape to adjust the bending that comes from the cornea so that in the end, the image is focused right on your retina.

Glass Helpers

There are vision problems other than astigmatism. Some people have corneas that bend light too strongly. Others have eyeballs that are too long. When either of those things occurs, the lens ends up focusing the image in front of the retina. As a result, the image on the retina is blurry. Since things that are up close need to have light bent a lot in order to put them in focus, people with this kind of problem can see things well up close, but they cannot see things very well when they are far away. As a result, we say these people are **nearsighted**. The medical term is **myopia** (my oh' pee uh).

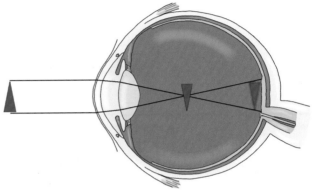

When you are nearsighted, the image is focused on a spot in front of the retina. What ends up on the retina is blurry.

When a person's cornea doesn't bend light well enough, or when the person's eye is too short, the lens focuses the image behind the retina. Once again, this causes the image on the retina to be blurry. People with this condition can see things that are far away well, but they cannot see things well up close. We call them **farsighted**. Medically, we say they have **hyperopia** (hi' puh roh' pee uh). Whether a person is nearsighted or farsighted, glasses (or contacts) can be made that correct how light is being bent so that the image ends up being focused on the retina.

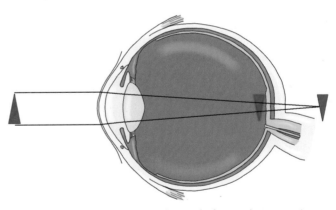

When you are farsighted, the image is focused on a spot behind the retina. What is on the retina is blurry.

Oftentimes, children do not realize they need glasses. After all, if they don't know what things are supposed to look like, how do they know that they are seeing incorrectly? That's why it is important to get your eyes checked regularly. A good eye test will tell you whether or not you are seeing things the way you should.

Try This!

Let's test your vision! On the last page of this lesson (page 193), there is a chart you can use to test your vision. You will need to have someone hold up that page 10 feet away from you. Typically, these charts are placed 20 feet away, but our chart is smaller, so we will place it 10 feet away. How many lines were you able to read? If you can read the line that is one up from the bottom, you have normal 20/20 vision.

Upside Down World

When you look at something, your eye gathers light from that image and focuses it onto the retina. As you saw a while back, that image is actually flipped upside down when it forms on your retina. The reason you don't actually see things upside down is that while you are an infant, you learn to link the image on the retina to the objects in the outside world. Since you spend time as an infant touching everything, you learn the correct orientation of everything, so you get used to flipping the image in your mind so you see things right side up. Interestingly, when an image is focused on your retina, it can stimulate the rods and cones there for a bit before fading. This is the reason animated cartoons can make things look like they are moving.

 Try This! To see how this works, take a stack of index cards and illustrate a series of stick figures walking or doing a cartwheel. Each index card should picture your figure moving a little further in the action than the previous card, with the last card showing the end of the action. When you're done with your illustrations, rapidly flip through the cards. Your stick figure will appear to move, because each time you flip to a new card, the preceding image is held briefly on your retina, and your brain melds them together if you flip the cards quickly enough.

If you put each of these drawings on a card and flip through them quickly, it will seem like the man is walking.

 Try This! Did you know that you have a **blind spot** in your eye where there are no rods or cones? It's true! This is the place where the optic nerve is located.

Below this paragraph, you see a circle and a plus sign. Have someone hold this page up in front of you a few feet away. Now, cover your left eye and look at the plus sign with your right eye. Next, have your helper slowly bring the book closer and closer. Stay focused on the plus sign. What happens to the circle?

You should have seen the circle disappear as your helper brought the book closer to you. That's because the circle eventually landed on your right eye's blind spot, and you could no longer see it. With both eyes open, this won't happen, because the circle won't hit both your eyes' blind spots at the same time. Your brain knows about your blind spots, so it uses the information from one eye to "fill in" the information that is missing from the other eye's blind spot.

Double Vision

God created both people and animals with two eyes. Most animals have their eyes positioned on the sides of their heads, and they look at the world through each eye separately. However, God designed the eyes of people (along with some animals) to work together so that we can locate objects more accurately. We call this **binocular vision**. It increases our depth perception (the ability to determine how far away something is) and our ability to judge the location and size of things.

A frog has two eyes, but it doesn't have binocular vision. As a result, it has poor depth perception.

Try This!

Sit in a chair and have someone hold a marker with a brightly colored lid at your eye level about 5 feet away from you. Now have that person slowly walk in an arc. Without turning your head, see how far along the arc you can follow the marker with your eyes. Now try this with one eye closed. What happened? Binocular vision not only increases your visual field, but it also helps you to judge distances.

Let's keep experimenting! For this activity, you and your partner each need a marker. Now, both of you stand. Have your partner hold his marker at about waist level (so if you poke your partner, you won't do too much damage). Close one eye, and try to touch your marker lid to your partner's marker lid. Now try with the other eye, and then with both eyes. What do you find? For more excitement, try playing catch with a Nerf ball while you keep one eye closed. Throw the ball gently, please!

Eye Will Protect You

God designed a lot of protective gear for your eyes. Some of the strongest bones in your body surround your eyes. This is so that if you get hit in the face near your eye, your bones will not crush your eye. Once, my son was playing golf and was hit by a golf ball right near his eye. It didn't do much damage because the special design of his skull protected his eye.

Do you know why you blink? Your eyelashes and eyelids blink to protect your eyes, preventing foreign objects from entering into them. But that's not all! You have special glands that constantly wash your eyes. They produce a wet substance designed to moisten, cleanse, and disinfect your eyes. We call them **tears**. God designed special tubes called **tear ducts** that lead from the corners of your eyes into your nose, so that the fluid has a place to escape.

When you begin to cry, there's too much fluid to pass through the tear duct and

This little girl has pinkeye. Although it is uncomfortable, it can be treated with antibiotic eye drops.

into your nose, so the extra fluid leaves your eye as a tear. God designed your eyes to tear up when anything gets in them – and also when you're really sad or really happy. Of course, some children can fake crying. Can you do that?

Have you ever known someone with **pinkeye**? The other name for this condition is **conjunctivitis** (kun jungk' tih vye' tis). The **conjunctiva** (kun' jungk tih' vuh) is a protective covering over the white part of the eye. This thin, clear membrane is another mucous membrane. Like the mucous membrane in your nose, it can become infected. When infected, it turns pink. It usually causes your eye to itch or burn. We call this pinkeye, and it is very contagious. Although annoying, it is not serious and can be treated with the proper medicine. The most important thing is to avoid giving it to others.

Eye Understand

Well, now you know the basic features of the eye. However, the eye is far more complex than even what you have learned here! God truly gave people a gift when He designed our eyes. How wonderful that we can look around us and see the beautiful world God created. Yet, even more wonderful is the fact that God also has eyes, and He is always watching over us! He makes us this promise: "*I will instruct you and teach you in the way which you should go; I will counsel you with My eye upon you*" (Psalm 32:8).

What Do You Remember?

What is the part of your nose that holds your olfactory cells called? How are odors received and transferred to the brain? Where are your taste buds found? What are the five taste sensations? What is the pinna? Why do ears make wax? Name the three bones in the middle ear. What are otoliths? What is the sclera? What is the pupil? What is the iris of an eye? What is the fovea? What cells enable you to see in color? What cells enable you to see in dim light? Name some of the ways God added special protection for your eyes.

Notebooking Activities

Record all that you learned about the five special senses in your notebook. Be sure to include illustrations. Also, make a diagram of the eye and label all its parts. Write about what each part does for the eye.

Experiment
Testing Taste

You did several experiments to test your sense of taste. You discovered that your sense of taste is dependent on your sense of smell. But is your sense of taste dependent on the sense of sight as well? Find a volunteer to help you do this experiment so you can find out!

You will need:

- A volunteer willing to taste foods while blindfolded
- A good blindfold for your volunteer's eyes
- A variety of foods with sweet, salty, bitter, sour, and umami (savory) tastes, such as:

Ice cream	Vinegar	Turkey
Banana	Salt	Lemon slices
Chocolate	Sugar	Honey
Milk	Coffee	Syrup
Orange slices	Bacon	Butter

- Straws for testing liquids
- Spoons for putting the food on the volunteer's tongue

1. Choose the foods you will use and place them in separate containers out of sight from your volunteer taster.
2. Blindfold your volunteer.
3. Before placing the food on the volunteer's tongue, have your volunteer hold his nose so he cannot smell it.
4. Begin by having your volunteer open his mouth and stick out his tongue.
5. Place each item on the volunteer's tongue. If you are using liquids, simply dip the straw in the liquid and a drop will adhere to the straw. Then, place the straw on the tongue, allowing the drop to move onto the tongue of the volunteer.
6. Ask your volunteer to identify the food based on taste alone.
7. Now repeat the experiment, but this time allow your volunteer to smell the food.
8. Keep a record of which foods your volunteer could identify with neither sight nor smell as well as which he could identify with smell but no sight.

What did you learn? Were there foods that your volunteer couldn't identify even when he was allowed to smell them? If you want, try this on other people. See if some people are more dependent on sight for taste than others.

Visual Acuity Chart - Approximate Snellen Scale
For Educational Purposes Only

20/200	**E**	200 ft / 61 m
20/100	**H N**	100 ft / 30.5
20/70	**D F N**	70 ft / 21.3 m
20/50	**P T X Z**	50 ft / 15.2 m
20/40	**U Z D T F**	40 ft / 12.2 m
20/30	**D F N P T H**	30 ft / 9.1 m
20/20	**P H U N T D Z**	20 ft / 6.1 m
20/15	**N P X T Z F H**	15 ft / 4.6 m

193

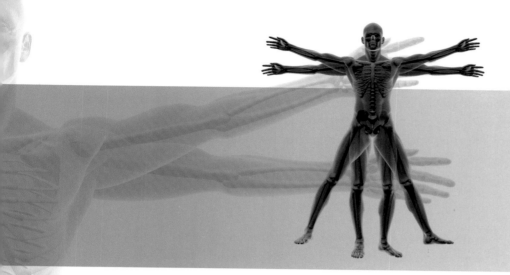

the
INTEGUMENTARY SYSTEM

Let's now cover everything we've learned so far in this book. Don't worry! I'm not saying we will review all that we've learned; I'm saying we need to cover it up! You see, one organ God created for mankind is so familiar we hardly think of it as an organ at all. It's the organ that covers you from head to toe – your skin! Skin is so familiar that we rarely give it a second thought, unless it's causing a problem. But did you know that skin is the largest organ in your body? It's true! In fact, your skin takes up about 15% of your body weight. So, if you weigh 60 pounds, nine of those pounds are from your skin alone!

Skin is truly a sensational structure that only God could have created. Your amazing skin is stretchy, water-resistant, flexible, and self-repairing. It helps to regulate your body temperature, prevent infection, protect you from the harmful rays of the sun, and keep you from dehydrating. It also produces an important vitamin. When you walk outside, your skin tells you immediately whether or

Your skin is the largest organ in your body.

not you are dressed appropriately for the weather. Because it's full of sensory neurons, your skin gives you a great deal of information from the world around you.

Do you remember from the very first part of this book that an organ is a group of tissues working together to carry out specific functions or jobs? Well, it's no different for the special group of tissues we call skin. There are several tissues that all work together to make what we call skin. It is made up of skin cells, blood vessels, nerves, and connective tissue. Together with your nails and hair, sweat glands, and oil glands, your skin makes up your **integumentary** (in teg' yoo men' tuh ree) **system**: a complex group of tissues working together to ensure your survival. Without it, you just wouldn't look quite right!

Stretch and Grow

Skin, nails, and hair all grow in a similar pattern. Learning this pattern will make learning about the integumentary system a lot easier. In a nutshell, hair, nails, and skin all grow only at the base – where they begin, underneath the surface. As new cells are formed at the base, older cells are pushed outward. As these older cells are pushed up and away from the base by the newly forming younger cells, the older cells die off. That may seem like it's the end of those old-timers, but it's not! In fact, as they die off, something fascinating happens – they harden. Once they harden, they become skin, nails, and hair. These are three important dead things you have on your body, wouldn't you agree?

Dearly Departed Hair

Let's look at hair as an example. I'm sure you've noticed when you get a haircut that your hair doesn't hurt or bleed where it gets trimmed. That's because the cells that make up the hair hanging off your head are dead. Since the cells are not alive, they have no nerves or blood supply. But your hair continues to grow, so surely it must be alive? Well, hairs are living, but only at the base. That's why it hurts when someone pulls your hair or yanks a strand out.

Look at the drawing on the right. Hair grows from a little "pocket" called a hair follicle (fol' ih kuhl), which contains living cells. The cells in a region called the matrix reproduce quickly, making new cells. The new cells are pushed outward by more newly-made cells. As the cells grow away from the base, they become progressively farther from the blood supply. Without a fresh supply of nutrients, they die, forming the shaft that you recognize as hair.

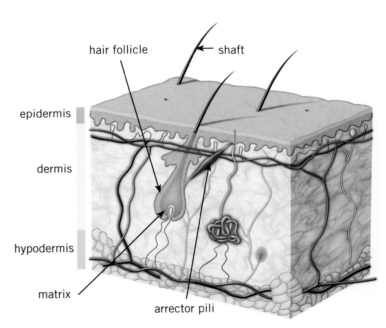

Skin has two layers (epidermis and dermis) that sit on top of the hypodermis.

A very similar process happens with your nails as well as your skin. Did you know that the outer layer of your skin consists of cells that have died? This is most obvious when your skin is very dry. Gentle scratching will cause bits of the outermost layer of dead cells to brush right off. But every day, all the time, you are shedding individual skin cells – or small clusters of skin cells – that have died. Frankly, a lot of the dust in your home is actually shed skin! Adults shed about 2 pounds of skin cells a year. Go look in your kitchen cupboard for packages weighing between one and two pounds, and you'll get an idea of how much dust one single person makes in a year! Now that you've got the basic principle, let's take a closer look at your skin.

Skin Stories

Your skin has two layers. The outermost layer is the **epidermis** (ep' ih dur' mis) – "epi" means upon, and "derm" means skin. Its main function is protection. Below the epidermis is the **dermis** (dur' mis). It contains a lot of connective tissue as well as motor nerves for the tiny **arrector** (uh rek' tor) **pili** muscles that make your hair "stand on end." It also contains blood vessels, hair follicles, oil and sweat glands, and sensory nerves for pressure, temperature, and touch. Below the dermis is the **hypodermis** (hi' poh dur' mis) – "hypo" means under – to which the dermis is securely attached.

The hypodermis is not a part of the skin. Rather, it contains **adipose tissue** (fat tissue), connective tissue (everything needs to connect somewhere), blood vessels, and nerves. This layer is like a wet suit. Do you know what a wet suit is? Well, divers and surfers wear wet suits to keep them warm when they are diving down deep or surfing in cold waters. The hypodermis helps to keep you warm, just like a wet suit. It also protects you if you fall or bump yourself.

Thick Skin

We say a man has "thick skin" if he doesn't get his feelings hurt easily. We call someone who gets easily offended "thin-skinned." Those are just expressions, but the outer layer of your skin does have different thicknesses, depending upon where it is on your body. Check out the skin covering different areas of your body. Feel your elbows and feet. Give them a squeeze. Now feel your face. Does it feel the same to squeeze your face as it does to squeeze your elbows and feet? Where do you think your skin is thinnest? This might make you blink a bit, but it's actually the thinnest on your eyelids! Where is it thickest? If you guessed the soles of your feet, you are right. Your palms also have thick skin. Why do you think God made this so?

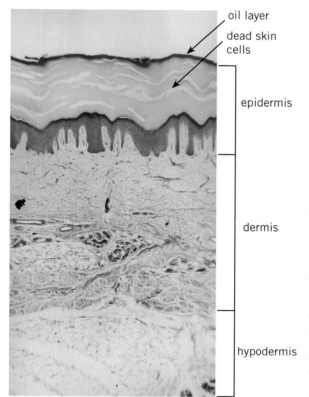

oil layer
dead skin cells
epidermis
dermis
hypodermis

This microscopic photo shows you the two layers of skin as well as part of the hypodermis.

Your Epidermis is Showing

The epidermal layer is the protective layer of your skin. It protects you from germs that are all around you, trying to get inside your body. It also protects all the layers beneath it. The epidermis is waterproof, keeping water out of your body and keeping your body's moisture locked inside. Without it, the cells beneath would all dry out, and you wouldn't be able to survive.

Do you realize that there are bacteria everywhere? Don't get frightened. It's always been that way, and you've survived just fine! But there really are bacteria all over your environment and all over your skin, all the time. The epidermis keeps these bacteria out of you, because God designed this layer of dead cells to interlock with one another, fitting together in such a way that nothing can easily penetrate. The only way bacteria can get inside your body is through an open door – a cut or wound on your skin. That's why washing cuts is so important. Once the bacteria get down to the moist tissues, they stand a much better chance of surviving and multiplying. Washing with soap and water as soon as possible helps destroy these invaders.

The living cells in the epidermal layer are called **epithelial** (ep' uh thee' lee uhl) **cells**. There are many layers of epithelial cells in the epidermis. At the bottom, near the blood vessels in the dermis, these epithelial cells are dividing and forming new skin. Here, these cells are plump. As they are pushed out by the newer cells, they move towards the surface. As they move towards the surface, they flatten out, making a substance called **keratin** (kehr' uh tin).

Keratin is a tough protein that is found in skin, hair, nails, the scales of fish and reptiles, the feathers of birds, and even the horns of rhinos and deer! So, keratin is always forming to make our skin tough. But the cells that produce keratin actually get so much keratin that they are poisoned by it. This leads to their death. The epithelial cells die from keratin toxicity as they near the surface of the skin. This is why the top cells of the skin are dead.

As the cycle goes on and on, the dead cells that flake off will be quickly replaced by new cells rising from the lower layers of the epidermis. This entire cycle, from cell division to cell death, takes three to four weeks. Cells remain on the surface for another week or so before being rubbed off by the activities of daily life. Even rolling over in bed will remove some!

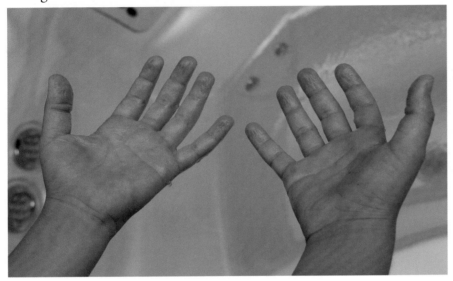

This boy's hands have been in water long enough for them to wrinkle.

Over the epidermis is a nice layer of oil that adds to the protection of your skin. However, if you spend too long in the shower, especially in hot water, the oil will wash away. You've probably noticed that your fingers and toes look bumpy when you've been in the tub or a swimming pool for a long time. This happens because water molecules seep into the epithelial cells when the oil is washed away, causing them to swell. The swelling is uneven, and the skin puckers – eventually you begin to look like a prune! After a while though, things return to normal.

Skin Deep

God loves variety, and He created many pleasing and beautiful colors for skin. Your skin is probably not the same color as your mom's. Your friends all have different colors of skin as well. These color differences come from pigments (color-giving chemicals) in the epithelial cells. There are two main pigments that influence the color of the skin, **melanin** (mel' uh nin) and **carotene** (kehr' uh teen). Carotene in the epidermis generally gives the skin a slightly yellow tone, while lots of melanin makes the skin brown, olive, or black. That's the only significant difference between people of different skin colors – they have different levels of pigments in their skin. While this is a noticeable difference, it is very minor compared to all the other differences that exist among people.

Carrots Please

You've probably heard of carotene, because it's found in orange vegetables, like squashes, sweet potatoes, and carrots. It's a yellowish-orange pigment, and it's what gives those veggies their characteristic color. Lots of babies love these vegetables because they're sweet, but if they eat too many, they turn a bit orange themselves! All that extra carotene travels in their bloodstream, so where their overlying skin is thin, you see the yellow hue shining through! This completely harmless color change is called **carotenemia** (kehr' uh tuh nee' mee uh), and it goes away if you reduce the amount of orange veggies the baby eats.

Even though what a baby eats can affect the color of her skin, the overall shade of your skin is determined by your DNA. If your DNA tells your skin cells to produce lots of melanin, it covers up the carotene, resulting in darker-colored skin. Fair (pale pink) skin does not have much carotene or much melanin. The pink comes from blood showing through the skin. Since there are 20 feet of very small blood vessels in each square inch of skin, it's easy to tell where the pink comes from!

Do you remember which foods contain vitamin A? Well, all the orange fruits and veggies have carotene in them, and your amazing body can turn carotene into vitamin A!

Melanin Melody

Melanin is produced by specialized cells called **melanocytes** (mel' uh no sytes') that are found deep in the epidermis. Everyone has about the same number of melanocytes, but skin color is determined by the amount of melanin your particular melanocytes make. This is controlled by your DNA, but it is affected by how much your skin is exposed to the sun. Melanin absorbs harmful rays that come from the sun. This protects the deeper layers of your skin. When your skin is exposed to the sun's rays, your melanocytes increase the number of melanin granules they are making, in an effort to better protect your body.

Do you have freckles on your face? They are caused by melanocytes. When there are tiny clusters of very active melanocytes, they make a freckle – an area on your skin that is darker than the rest. I'm sure you've seen people with red hair. Have you ever noticed that redheads tend to have orange freckles? This occurs because there are actually two types of melanin. Most people have mainly dark brown melanin, but some people have red-orange melanin.

This girl's red hair and red-orange freckles come from a specific kind of melanin.

Albinos are people or animals that cannot manufacture melanin. In the most severe cases, there is no melanin made at all. Hair, skin, and eyes are nearly white, causing the affected person or animal to stand out in a crowd. For people, not having the protective effects of the melanin makes cancers that come from sun exposure extremely common. In the animal world, albinos don't live long, since it's so easy for predators to spot the animal deprived of its usual camouflage.

Do you remember that you need plenty of vitamin D to make strong and healthy bones? Do you also remember that your body makes it from being exposed to the sun? Your skin uses certain rays from the sun to modify a form of vitamin D your body cannot use to a form that your body can use, after a trip through the liver and kidneys for some final modifications. So a bit of exposure to the sun is important, but too much isn't good for you. About 20 minutes a day of sun exposure in a swimsuit (more if you only expose your arms and legs) during midday provides the right

This South American family has one child that was born without the ability to make melanin in her skin cells. Even her eyes are without pigment, making them appear pink. We call this albinism. She will still be able to live a long, healthy life, as albinism does not affect a child's health with the exception of lack of pigment in her retina which will result in poor vision.

amount of sunlight for vitamin D to be created inside your body. After that, you're just heading for wrinkles and skin cancer, so be sure to wear a hat, some protective clothing, and some sunscreen if you are staying in the sun for a long time. If you do not get enough sun exposure, you need to get vitamin D from your diet or from vitamin pills.

Why, you may wonder, is it dangerous to get too much sun? Well, the sun produces a lot of different kinds of light, including very high-energy light called **ultraviolet** (ul' truh vye' uh lit) light (sometimes abbreviated UV light). You cannot see this light, but it has so much energy that it can actually kill skin cells! Now, of course God knows all about the sun's ultraviolet light, so He designed your body to be able to deal with some of it. However, too much of it can cause too much cell death, and it can also lead to skin cancer. Scientists have discovered that the more a person is exposed to the sun without sunscreen, the more likely it is that the person will develop skin cancer. Also, people with less melanin in their skin are more likely to get skin cancer, since melanin absorbs UV light before it does damage. These people need sunscreen even more than others. Conversely, people with darker skin (more melanin) need more sun exposure to make the vitamin D that they need.

Tell someone all you have learned so far about the skin.

The Dermis

As you have learned, the epidermis is made up of epithelial cells. The dermis, on the other hand, contains connective tissue as well as special structures like hair follicles, sweat and oil glands, blood vessels, and nerves. Because of this, the dermis is much thicker than the epidermis. The connective tissue within the dermis is what makes skin stretchy and elastic, allowing you to move freely without needing baggy skin to permit free movement. The most common proteins making up the connective tissue are collagen and **elastin** (ih las' tin). You may have guessed that elastin makes skin – elastic! These amazing proteins allow your skin to stretch and recoil as you move about. Wrinkles occur as the body ages, because the skin becomes less elastic (as well as drier and somewhat thinner). Too much exposure to the sun speeds up this process.

 Look at the skin at your elbow. Now bend and straighten your arm while feeling the skin at your elbow. Now, feel the skin on your mom's elbow while she bends and straightens her arm. Which is more elastic: your skin or your mom's? You need extra skin at your elbow because of the huge range of motion your arm requires in that location. In young children, the skin is very elastic and not as wrinkled. As you age, the skin loses elasticity.

Bruising

Have you ever hit your leg really hard against something? If so, you probably got a bruise where you were hit. A bruise happens when damage to the skin causes a blood vessel (usually several blood vessels) to leak. The red blood cells spread out into the dermis until clotting happens. Then, over time, white blood cells arrive to clean up the mess. It can take up to a couple of weeks for the red blood cells in the tissue to be completely broken down and removed by the white blood cells. As the cleanup process continues, it can create a wide variety of colors in the skin as the hemoglobin from the dead red blood cells slowly breaks down.

You may be wondering why, if your outer layer of skin is dead, it isn't dry and scaly all the time. Well, God provided us with oil glands. They are made of epithelial tissue but are located down in the dermis. They are called **sebaceous** (sih' bay shus) glands, and they are like built-in skin lotion dispensers! They produce an oily substance we call **sebum** (see' bum), which coats the skin and hair, keeping them smooth and supple. Like most oils, it helps repel water. But, just like everything God makes, sebum is better than any skin lotion you could buy. This special oil actually kills many kinds of germs trying to invade your body.

Sebaceous glands are present in the dermis all over your body, but on your face and scalp there are many more per square inch than anywhere else. During adolescence (when kids are growing

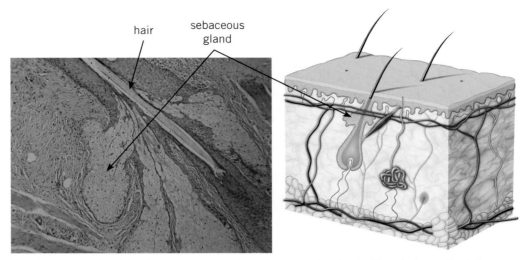

hair

sebaceous gland

The drawing on the right shows you where the sebaceous glands are. The photo on the left shows you what they look like under a microscope.

into adults) these glands tend to be very active for a few years, often causing extra oil on the face and oily hair. The hair gets oily because nearly all the sebaceous glands on your body empty into a hair follicle. In most places on your skin, the hair coming from the hair follicle associated with the sebaceous gland is so tiny you don't even notice the hair.

Bursting Blisters

Have you ever had a blister? It can happen when your shoes don't fit properly or perhaps when you touch a hot pan. Either way, the blister separates the epidermis from the dermis. Interestingly, until the blister pops, the raised layer doesn't hurt, because the nerves in the epidermis have been destroyed. When the blister pops however, you'll notice that the skin beneath is moist and very tender to even the lightest touch. This is because there are a lot of sensory nerve endings in the dermis, including pain sensors. You will learn about them in a little while.

Blisters are usually filled with fluid, some of which comes from your **sweat glands**. They produce sweat, which is released from pores on the surface of the skin. Every day, all the time, you are losing a very small amount of water this way, without even noticing it. Along with the water, some salt and a few other chemicals are being lost.

sweat duct

sweat pore

sweat gland

sweat gland

How long did it take for your blister to heal? God knew your skin would need frequent healing, so He provided a special healing mechanism within your body for when small areas of skin need healing. The hair follicles, deep within the dermis, are lined with epithelial cells. These cells can migrate outwards from the hair follicle and replace large areas of epidermis if need be.

Sweat glands in the dermis produce sweat that is then poured out on the skin through tiny sweat pores. When a blister forms, the sweat duct that carries sweat to the pore is damaged, and it releases sweat in the blister.

Don't Sweat It

Most of the time you only notice sweating when you're hot, because that's when you produce the most sweat. If it's extremely hot out and you're exercising, it's possible to lose up to 42 cups (10 liters) of water and up to two tablespoons of salt in one day. Why does your body get rid of water and salt this way? It sweats in order to cool you down. While the sweat sits on your skin, it begins to evaporate into the surrounding environment. This takes energy, and the sweat gets that energy from the skin. When it evaporates then, it takes energy (heat) away from your skin, cooling it. When the sweat cannot evaporate easily (like when it's muggy out), you can overheat and become quite ill if you continue exercising for too long or don't get to a cool shaded area.

Try This!

Go outside on a hot day, taking a very damp cloth with you. Slightly moisten one arm with the cloth and then concentrate on your arms. Which one feels cooler? Blow on the damp arm. Does it feel even cooler? The blowing speeds the evaporation, so it should feel cooler.

Heat Exhaustion

Heat exhaustion can lead to a serious illness called heat stroke if it's not treated. It's important for you to be able to recognize the signs of heat exhaustion so you can help someone before it develops into heatstroke. Heat exhaustion occurs when someone is exposed to high temperatures for a long time without regularly replacing lost salt and water. Elderly people can get heat exhaustion more easily than younger people.

Heat exhaustion can lead to heatstroke, so if you get hot while playing, take some time to cool off in the shade.

The symptoms of heat exhaustion are heavy sweating, pale skin, muscle cramps, tiredness, weakness, dizziness, headache, nausea or vomiting, and fainting. People experiencing heat exhaustion may have cool, moist skin and a very quick pulse. Also, their breathing will be fast and shallow. If you suspect someone is experiencing heat exhaustion, get the person out of the sun and into a shady or air conditioned area. Give him a cool drink. Remove excessive clothing and place cool cloths on his head and skin. Call for medical help if he doesn't feel better right away.

Thermostat

In addition to the sweat glands, the blood vessels in the dermis help with body temperature control. When your temperature rises a bit above normal, the blood vessels in the dermis widen to release heat through the skin. If it's cold out and your body senses your core temperature is drifting downwards a bit, those same blood vessels will constrict (get smaller), keeping the warm blood more towards the center of your body.

Hair Controls

Even the hairs on your body help with temperature control. If you're cold, you feel better when you cover more of your skin. This is because the covering (tee shirt or coat), traps a layer of warm air near your body. When you're cold, you may have noticed you get goose bumps. That's because the tiny muscles attached to the hairs in your dermis (the arrector pili muscles) tighten, causing the hairs to stand up. This causes the surrounding skin to mound up slightly, making the goose bump. This is all under the control of the autonomic nervous system. By going through this motion, the hairs trap a layer of warmer air near your skin. This isn't quite as effective as a sweater, but it does help.

Very Hairy

All over your body, you have millions of hair follicles. Some of your hair is very, very fine, so you don't generally notice it.

Try This! Get a mirror and look very closely at your face. If the light hits it just right, you can see that you have hair all over it!

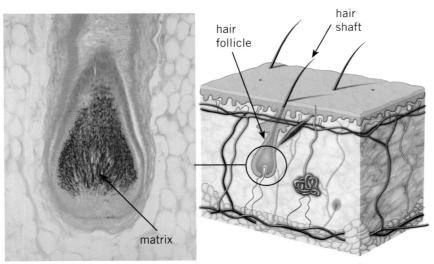

hair follicle

hair shaft

matrix

The photo on the left shows what the matrix of a hair follicle looks like under a microscope. The colors are the result of chemical stains that make things easier to see.

As you can see from the image on the left, the hair follicle is a deep pit of epidermal tissue within the dermis, where it has access to an excellent blood supply. The follicle is lined with epithelial cells. At the base of the follicle is the **matrix**, the site of hair growth. The hair cells in the matrix divide, creating new hair cells and pushing the older cells outward. As they move outward, they produce so much keratin that they die, much like your skin. The dead cells make up the **shaft** of the hair. Now the keratin made within hair cells is slightly different from the skin's keratin. It is called **hard keratin**, a good name when you feel your hair compared to your skin.

Although all hairs on your body grow at more or less the same rate, some of your hairs are long and some aren't. That's because hair roots go through growth and rest cycles. The hair grows during the growth cycle. During the rest cycle, the hair remains in place, and the hair follicle rests from the work of hair production. After the rest cycle, new hair starts to grow, pushing the old hair out. Then the entire cycle repeats. The longer the growth cycle, the longer the hair will be.

Every day of your life about 90% of your hair follicles are in the growth phase, so most of your hairs are growing most of the time. The hairs on your head grow for about 3-4 years and then rest for 1-2 years.

Then they fall out. Because they have a long growth phase, they can get very long. Even though you may not have noticed, some of your hair falls out every single day. This is because there are so many hairs on your head (probably over 100,000), that some of them come out of the rest cycle every day. Your eyelashes grow for only 3-4 months before the hair root takes a rest. But eyelashes have a very long rest phase, so you probably don't notice a lot of them falling out.

Try This!

Grab a handful of your hair and pull to see if any hair comes out easily. If your hair is dead, why does it hurt when it's pulled? Because the follicle is alive and is well supplied with nerves. Now look to see how many hairs came out. The hairs that pulled out were finished with the growth and rest cycles and were ready to come out. If you've recently brushed your hair, it may be more difficult to find hair that is ready to come out, as it was probably released into your brush this morning. Save one strand of hair for the next section.

Try This!

Close your eyes and ask someone to gently brush your arm, just above your skin. He shouldn't actually touch your skin, but simply brush his hand right above it. You'll notice you can feel your hairs move! That's because you have nerves wrapped around your hair follicles, and those nerves tell you when your hairs move.

Layered Hair

Hair itself has three layers: the **cuticle** (kyoo' tih kuhl), the **cortex** (kor' teks), and the **medulla** (mih doo' luh). All three layers contain hard keratin. The outermost layer, called the cuticle, is made of hard, overlapping cells, like the shingles on a roof. In fact, the cuticle makes the hair water-resistant, like the shingles on your roof keep water from entering your home. The cells of the cuticle layer are clear, even though the hair is colored. The cuticle allows stretching of the hair without breaking. Hair that's extremely soft is called "baby-fine hair." It has fewer overlapping layers of cuticle cells than does thicker hair. The thicker the hair, the more cuticle it has.

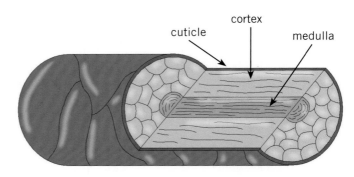

Hair is composed of three layers. The cuticle is mostly transparent, so the color comes mostly from the cortex.

Try This!

Take the strand of hair you saved and very carefully pull on it. Do you see how it stretches before it breaks? This is because of the stretchy cuticle layer.

The color of your hair is found in the cortex layer. Remember, the cuticle is clear, so you can see the color of the cortex through the cuticle. Melanin and similar pigments are taken up by the cells of the cortex, and the mix of pigments determines the color of the hair. As a person ages, the amount of pigment in the hair decreases, causing it to appear white. Gray hair is actually a mixture of pigmented and white hairs.

The inside layer of hair is called the medulla. It is made of loosely connected cells. In fact, in many hairs, the medulla is not continuous, which means there is air between clumps of medulla cells. Also, not all hairs have a medulla. Fine hair and naturally-blonde hair rarely have a medulla.

Straight or Curly

Whether you have straight or curly hair all depends on the shape of your hair follicles. If your hair follicles are round, you'll have straight hair, because the hair can grow straight out of the round follicle. It doesn't need to twist and turn to get out of there. Wavy hair is produced by oval follicles. The round hair has to wiggle out, becoming wavy as it does. Curly hair is produced by flatter, oval-shaped follicles. People with hair that grows in tight ringlets have very flat follicles. Which kinds of follicles do you have?

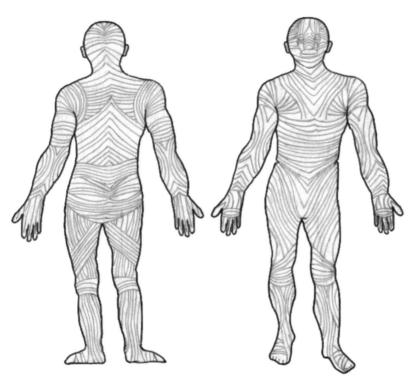

Cleavage lines in the skin affect how wounds heal.

Dermal Indentions

Have you or someone you know ever had a sharp object, like a nail, puncture the skin? How did it heal? You probably noticed that it didn't heal in the shape of the object that pierced the skin. When healing occurs in the skin, it typically leaves a slit for a scar. This is because skin has patterns deep within the dermis. These dermal patterns are like deep, thin canyons that run all over your skin in long paths. Scientists call these dermal indentions **cleavage lines**. They have maps of these lines, like the ones shown on the left. If a person gets injured and the injury runs *parallel* to a cleavage line, when the injury heals, you will probably not be able to see a scar. However, if the injury cuts *across* a cleavage line, usually a scar will be visible. You can see these cleavage lines most easily if you look at the palm of your hand. Do you see the deep grooves? Those are valleys and canyons in the dermis. Many cleavage lines are hard to see, but you can bring them out if you try. If your brother has on a swimsuit, look at his back while asking him to stick his chest out and pull his shoulders back. Amazing, huh?

My son was bitten on the face by a dog. Thankfully, the dog bite followed a cleavage line. Today, the scar isn't visible because of this. If you ever have an injury to your face that needs stitches, you should have a plastic surgeon do the stitching, because these surgeons are trained to stitch along the cleavage lines to minimize scarring. Other doctors might not be as careful.

Look at the tips of your fingers. Do you see some dermal indentions there? Your fingerprints and toe prints aren't straight; they are whirly and swirly and are solely yours! God gave each person a set of fingerprints like no other fingerprints in the world. You are so special that no one who has ever lived, and no one who will ever be born, has the same set of fingerprints as you. Even identical twins have different fingerprints. In fact, it is because of this that forensic scientists are able to find missing persons and solve crimes – by locating fingerprints. Even if you cut your finger deeply, your fingerprint will grow back exactly as it has always been. This is because fingerprints are created deep in the dermis. These special whirly patterns give your fingers traction so things don't slip out of your hands as easily. They also give your feet traction when you're walking barefoot.

The ridges in your dermis form fingerprints, which are unique.

Even though we usually can't see them, we actually leave our fingerprints almost everywhere we go because the oil and sweat glands cause our fingertips to be slightly greasy. When a crime has occurred, scientists will find and lift up fingerprints so they can be taken to a lab where computers match them to individuals. Chemicals are often used to take fingerprints off many surfaces, but one way that forensic scientists locate and lift fingerprints is to "dust" for them. They lightly dust an area that might have prints, and the oil left behind from the person's fingers traps the dust in the pattern of the fingerprint. The investigator can then pick up the dust pattern with tape, or she can take a picture of the pattern. Either way, when the pattern is analyzed, it will indicate only one person in the entire world!

Happy Hypodermis

The hypodermis is not a part of the skin, but since it is underneath the skin, we should talk about it. It is also known as **subcutaneous** (sub' kyoo tayn' ee us) tissue. Because "hypo" means under, hypodermis simply means under the skin. You may remember that much of the hypodermis contains many fat cells. It also has loose connective tissue, which connects it to the dermis. In fact, the hypodermis keeps your skin from sliding around on your body.

Try This!

Remember feeling your elbow earlier? Well, let's feel it again. Extend your arm out and feel that skin loosely moving around. Now feel the skin on your knuckle. Compare that to the skin on your arms and legs. Which parts do you believe have more subcutaneous tissue? The skin on your elbows and knuckles seems to slide around quite a bit. That's because it has very little hypodermis tissue beneath it. The dermis is just loosely attached to the bones in the joint. If all the skin on your body were like that, you would be sporting a lot of wrinkles!

Sensing General Senses

We told you earlier that the dermis contains nerves. Look at the diagram on the next page to see how many there are. We covered the special senses that God gave you in the previous lesson; now we'll talk about the general senses – those distributed all over the body. The reason they are distributed all over the body is because they're found in the dermis, as well as in the muscles and tendons.

Your skin is sensitive – very sensitive! In fact, there are all kinds of sensors in your skin. Let's peek at a few hiding under there. Close to the surface, right in the epidermis, you can find free nerve endings. **Free**

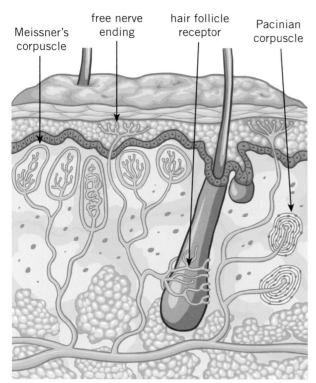

Meissner's corpuscle free nerve ending hair follicle receptor Pacinian corpuscle

The dermis is filled with nerves.

nerve endings are basically dendrites that are stimulated by heat, cold, pain, itch, or skin movement. Can you find them in the drawing? There are more of these in certain areas of the body than in others. For instance, your face is far more sensitive to light touch than is your lower back. The free nerve endings that sense heat are not really temperature sensors. Instead, they sense changes in temperature. When you jump into a cold pool on a hot day, for example, the free nerve endings in your epidermis register the change in temperature.

Try This!

To see your heat sensors in action, get three bowls of water: one cold, one lukewarm, and one hot (but not so hot that you'll burn yourself). Place a finger from one hand into the cold water, and place a finger from the other hand into the hot water. Wait about ten seconds so they become accustomed to the surrounding temperature. Now put both fingers into the lukewarm water. What do you notice? The finger from the cold water will sense hot, while the finger from the hot water will sense cold. That's because your nerves sense the change in temperature, not the temperature itself.

This is also the reason people from different climates don't feel the weather in quite the same way when they are in a different region. For example, my aunt came from Colorado to visit us in Texas one winter. The winter temperatures were in the 40's and 50's, and we were all quite cold – wearing heavy winter coats and shaking when we were outdoors. My aunt, however, wore short sleeves and had to sleep with the window open. She found the 40 to 50 degree temperatures to be warm and refreshing. We thought she was insane!

Let's look at a few more of those sensors in your skin. Towards the upper edge of the dermis are the **Meissner's** (myz' nerz) **corpuscles** (kor' puh sulz). These tiny "balls" give us the ability to tell the distance between two objects that are close together. Like other senses, there are more of these in certain areas (like your face and fingertips) than there are elsewhere in your body. You will do an experiment about this at the end of the lesson.

God sure designed a lot of different kinds of sensory neurons, didn't He? Do you remember how sensitive your hairs are to even light touch? That's because of the **hair follicle receptors** that are wrapped around each hair follicle. Even deeper, you'll see the **Pacinian** (puh sin' ee uhn) **corpuscles**. Find those in the drawing above. These special sensors can feel vibration and pressure.

Loop one end of a rubber band around a stationary object. Holding the band with a thumb and index finger, pull it taut. Now pluck it like it was a guitar string. Can you feel the vibration in the index finger that's pulling on the rubber band? Your Pacinian corpuscles are responsible for that. You could probably hear the "twang" as well. Sound waves result from vibration spreading through the air.

Some areas of your body are more sensitive to pressure than others. You can test this by applying the same amount of pressure to different areas of your body with your finger tip. Which areas are more sensitive? Did you guess your face and fingertips? Why is this? It's because in these two areas, there are more pressure sensors per square centimeter of skin.

Let's go down even deeper still! Way down in the lower part of the dermis are the sensors for even greater pressure and for stretch. These are sensors that are most likely to be stimulated only during significant pressure and stretching. They'll let you know if you're being pinched or if your skin is being stretched.

Before moving on to nails, explain what you have learned so far in your own words.

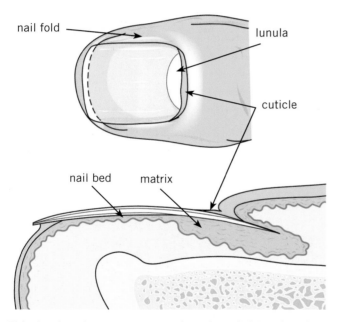

This drawing shows you the top view of a nail (above) and the side view (below).

Nails

The nail lies on the nail bed and is almost completely dead. Look at your nails. There they lie! Do they need a cleaning? If you're like most kids, they probably do. Look closely at the sides of each nail. Your nails are held in place on the sides by folds of skin called the nail folds.

The base of the nail (where it grows out of your finger) is the cuticle. It provides a brilliantly-engineered protective barrier, keeping foreign objects out of the matrix, which is the area that produces new cells for the nail. Look very closely at your nail just beyond the cuticle. What do you see? You should see a little white half moon. This lunar-shaped pigment is called the **lunula** (loon' yuh luh), or little moon. This white semicircular area is the part of the nail matrix that you can see. The rest of the nail bed appears pink because of all the blood vessels beneath the nail.

Nails are formed by cells dividing in the matrix. After the cells divide, they begin making hard keratin. The newer cells push the older, harder cells away from the root, and the cells soon die from keratin overdose, forming the nail that you end up clipping regularly. So you see, skin, hair, and nails all grow in a similar way. Your nails grow about 3 mm a month.

So, why do you have nails? What is their purpose? Sometimes they come in handy when trying to pick off a band aid or some sticky material. But what is their real God-given purpose? They certainly weren't created just to paint and look pretty on girls, were they? No. Nails serve to protect our fingers and toes, which are often bumped against things. Have you ever seen a building in progress where everyone has to wear a hard hat? The hard hat protects the workers' and visitors' heads. Your fingernails are like hard hats for your fingers! It's very painful to lose a fingernail. My daughter loses toenails now and again because she dances in pointe (ballet) shoes quite often. This causes the little hard hats on her toes to become quite battered. My coauthor, Dr. Ryan, has a young nephew who recently lost his fingernail after his finger was pinched in a door. As the new nail is growing in, his fingertip is experiencing a lot of minor trauma and is slightly swollen and red much of the time.

Your nails also make it easier for you to handle different things. They provide a firm surface for your fingertip tissues to push against as you grasp objects, making it easier to pick them up. Imagine trying to pick up a coin that fell on a flat floor if you didn't have fingernails! Fingernails also help you scratch your skin. Scratching removes irritants, increases blood flow, and relieves itchiness.

What Do You Remember?

What is the system called that includes your skin, nails, and hair? How do skin, nails, and hair grow? What are the two layers of the skin? What is the layer of tissue underneath those two layers? What two important pigments are mostly responsible for skin color? What do sebaceous glands make? What does sweat do? What are the three layers of a hair? Which layer is sometimes missing? Which layer of skin is responsible for making your fingerprints? In both hair and nails, what is the name of the region in which new cells are made?

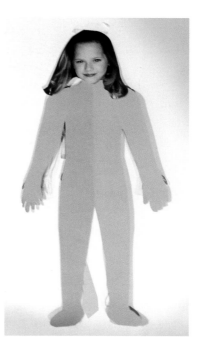

Personal Person Project

You will now complete your Personal Person by adding the final layer - the skin. You began with a skin layer on the bottom, now you'll finish up with a layer on top, to cover up all those organs inside your Personal Person. Using a piece of paper that is colored about the same as your skin, cut out the same body shape as you did in Lesson 1. Position this layer of skin on top and secure it at the feet, so that when you lift it, you will fold it downward. This will enable all the organs under the head to be visible when you lift the head. If you have the *Anatomy Notebooking Journal*, you will find a template to use for this layer of your Personal Person. Congratulations! You now know where every major organ in your body is located and have built a model to help you remember!

Notebooking Activity

After writing down all that you remember about the integumentary system for your notebook, make an illustration diagramming the layers of the skin. Include hair follicles, nerves, and other structures you might find in the dermal layer. Use the drawings on pages 196 and 207 as guides.

Although they are unique and different from everyone else's fingerprints, all fingerprints follow typical patterns. Some people in the same family have the same patterns. Others don't. Let's take imprints of your fingerprints for your notebook and find out which kind you have! First study the images on the top of the next page with the different fingerprint patterns. Just by looking at your fingers, are you able to guess which pattern they follow? Now let's actually take your fingerprints and see if you were right.

Arch
The ridges form a bump near the center, but most of the other ridges don't swirl around it.

Loop
There is a loop that generally tilts to one side, and most of the other ridges follow the loop's pattern.

Whorl
The central ridges form a spiral.

You will need:

- A sheet of paper with ten boxes labeled (one for each finger), similar to the one below
- A stamp pad

Practice making fingerprints on a sheet of scratch paper until you determine the right amount of ink and the right amount of pressure to put on the paper. You will want to make sure you do not smudge the ink on your paper. Once you feel confident, place a fingerprint on each box using the finger listed.

Compare your fingerprints to the patterns on the top of the page. Were you correct in your guesses?

Left Pinky	Left Ring	Left Middle	Left Index	Left Thumb

Right Thumb	Right Index	Right Middle	Right Ring	Right Pinky

Project
Braille Challenge

People who are blind are able to read thanks to a marvelous invention by a man named Louis Braille. This system of reading uses the highly sensitive tips of the fingers in order to feel raised dots that represent each letter. You will make your own set of braille letters and challenge yourself to learn the braille alphabet not only by sight, but by touch!

Make a photocopy of the braille chart here (you can print one found on the course website I told you about in the introduction to this book, or if you have the *Anatomy Notebooking Journal*, one is there for you to use). Then, using white school glue, place a dot of glue on each letter's dots and allow the glue to dry thoroughly. This will turn the dots on the page into bumps you can feel with your fingers. You can do the next experiment while you wait for the glue to dry. It should be ready in a day or two.

After your dots have dried, begin to feel the letters, teaching yourself which dots represent each letter. Once you think you know the alphabet, cut out each letter and, with your eyes closed, see if you can put them back in alphabetical order using just your sense of touch.

Braille Alphabet

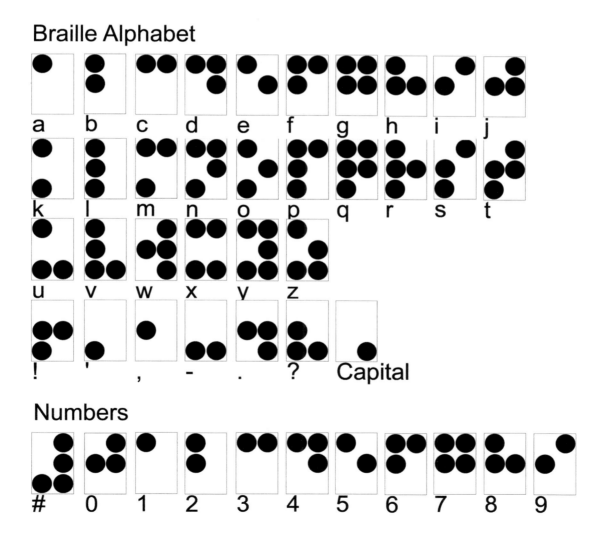

Experiment
Sensing Sensitivity

As you have learned, your skin is equipped with many sensory nerves. But which parts of your body have the most nerves in the skin? Can you guess? Use a Scientific Speculation Sheet and try this experiment with another person to see which parts of the body are the most sensitive. Also, if you wish, you can expand the experiment to see if people are more or less sensitive as they age.

You will need:

- A volunteer
- Five large paper clips
- A ruler or tape measure
- A chart with all the body parts you intend to test

Here are the body parts you will test:

- Back
- Back of arm above elbow
- Back of arm below elbow
- Stomach
- Top of foot
- Bottom of foot
- Back of hand
- Palm of hand
- Fingertips
- Forehead
- Cheek
- Nose
- Upper lip
- Tongue

Make a hypothesis about which body parts will be the most sensitive.

1. Unbend a paperclip and form it into an arch. For the first paper clip, position the two points of the paperclip 3 inches apart. Use the ruler to make sure they are 3 inches apart.
2. Form the second paper clip into an arch, with the points 2 inches apart.
3. Form the third paper clip into an arch, with the points 1 inch apart.
4. Form the fourth paper clip into an arch, with the points ½ inch apart.
5. Form the fifth paper clip into an arch, with the points ¼ inch apart.
6. To do the test, start with the widest paper clip that will fit in the area being tested. Have the volunteer look away, and touch his skin with both ends of the paper clip at the same time.
7. Ask him whether or not he can tell that there are two different points touching him. If he can, repeat the test with the next widest paper clip. Continue until your volunteer can no longer tell there are two different things touching him or until you reach the fifth paper clip.

The parts of the body where your volunteer was able to determine that two things were touching him when you used the fifth paper clip are the most sensitive to touch. Was your hypothesis correct? This kind of test is called a **two-point discrimination test**, since you are asking your volunteer to discriminate between two points that are touching his skin.

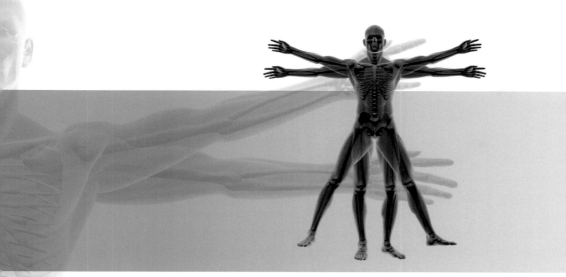

the
LYMPHATIC AND
IMMUNE SYSTEMS

Have you ever thought about how closely God watches over you? As amazing and miraculous as it may seem, God is always there watching everything you do, knowing the thoughts you have, and understanding every choice you make. The Bible tells us, "*All the days ordained for me were written in your book before one of them came to be. How precious to me are your thoughts, O God! How vast is the sum of them!*" (Psalm 139:16-17).

Since you are God's own child that He has known since even before you were born, and since He thinks about you all the time, you can imagine how much time He spends making sure you are safe and healthy! The Bible tells us, "*But the Lord is faithful, and He will strengthen and protect you from the evil one*" (2 Thessalonians 3:3).

There are many ways in which God shows His care for you. Often, He intervenes in your life through your prayers and keeps you safe from the evil around you. God also designed an amazing, intricate system inside your body that watches over you and fights for you if an enemy invades your body. This system is called the

Your body has all sorts of "protectors" that "beat down" the invaders that are trying to infect you.

lymphatic (lim fat' ik) **system**, and it protects you from invaders and gives you **immunity** to many infections. The lymphatic system, with its ability to provide immunity, is so complicated that even a basic understanding of it – which you will get in this lesson – convinces us that it could not have just happened by chance. It is so sophisticated and highly developed in design that there is no way it could exist without a Designer – our Lord and God!

As you learn about the lymphatic organs and immunity during this lesson, consider how amazing it is that your body works in such specific ways to protect you from harmful invaders: bacteria, viruses, worms, and other parasites. These organs are active all the time, working throughout your body, keeping it under close surveillance, like the cameras set up around the stores at a mall. Once a cell (or even part of a cell) that doesn't belong in your body is detected, a virtual army of your own cells is sent forth to isolate, attack, and destroy the threat! In order for this sophisticated surveillance system to function efficiently, it needs to have monitors all over your body. The different parts of the defense system also need to be able to communicate with each other. It may be tough to remember everything you will learn about the lymphatic system and immunity, but just remember this: only God could have designed such a specialized system – a system of tiny, living cells that have intricate war plans that they sometimes change so they can adjust to different situations. I pray that by the end of this lesson you will know how very much God cares for you, and you will give Him glory for your lymphatic system! Let's begin!

The Bad Guys

We'll start this lesson by discussing the bad guys: germs. Germs are microscopic things that enter your body and can potentially cause harm. Scientists call harmful germs **pathogens** (path' uh junz). When a pathogen gets inside your body and begins to multiply – making more of itself – you have an infectious disease. That sounds a lot scarier than it is. It just means there's an infection going on inside you.

An infection that only infects a small area (such as a splinter in your foot) is called a **localized infection**. Local means "restricted to one area." Once an infection is present and begins to spread throughout your body, it is called a **systemic** (sih stem' ik) **infection**. This is because it affects many

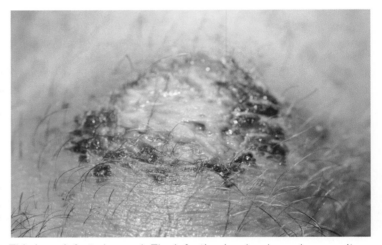

This is an infected wound. The infection is a local one, because it exists only in this wound – not in the entire body.

of your body's systems. Think about how you feel when you get the flu, which is a systemic infection. It makes you feel sick all over. Your muscles are sore, your throat hurts, and you cough a lot. Sometimes your digestive system reacts with vomiting and diarrhea. You usually run a fever, which makes your entire body hot. The systemic infection known as the flu is not fun to have!

Once, my son had a local infection from a dog bite to his face. The doctor stitched it up nicely. Yet, within a week, that local infection turned into a systemic infection. Because so many bacteria had penetrated deep into his tissue, he grew very ill; without the proper medicine, he might have died. We'll discuss what kinds of medicine can help against infection later on in the lesson.

Pathological Parasites

Parasites are living creatures that use your body to help them live and, in the process, make you sick. There are many different kinds of parasites. Let's look at some of the more common parasites that can infect people.

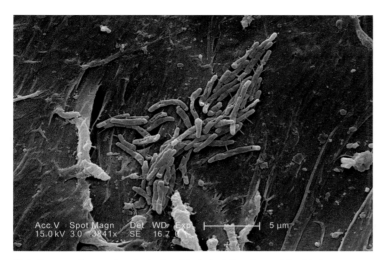

This image of **mycobacteria** (my' koh bak tear' ee uh) which cause eye, bone, skin, and lung infections, has been magnified by more than 3,500x. The colors are not real. They were added in order to make the bacteria (purple) easier to see.

Bad Bacteria

Bacteria (bak tear' ee uh) are one-celled organisms that are found nearly everywhere! Why are they so numerous? Well, because if conditions are right, they are able to multiply – making more and more bacteria – with incredible speed! Although some bacteria can make you very sick, it's important to remember that others don't cause any harm at all. Some bacteria live on your skin and in your body and don't hurt you one way or the other. Some bacteria even help you! Do you remember learning in the digestive system lesson that the bacteria living in your intestines make vitamin K for you? These are good guys, even though they are bacteria. The bacteria that live inside you and make you sick are the parasites. The ones that live inside you and help you are not.

Freaky Fungi

Fungi (fun' jye) are organisms that can also make you sick. Some infect the inside of your body. *Candida* (kan' dih duh) is a fungus (singular of fungi) that lives inside almost everyone. Your lymphatic system usually keeps it under control, but if it gets out of control, it can cause itching, rashes, and difficulty in swallowing. Some fungi infect your skin on the outside of your body. Athlete's foot, for example, is a burning, itching skin infection caused by the *Tinea* (tin' ee uh) fungus. These two fungal infections are easy to treat with medicine.

Wonky Worms

It's pretty weird to think about this, but it is possible for worms to infect your body as well. While these kinds of infections are much more common in undeveloped regions of the world like remote African villages, they can happen anywhere. For example, a microscopic worm called *Trichinella* (trik' uh nel' uh) can be found in undercooked pork. If you eat undercooked pork, the worm could infect you, causing the disease trichinosis (trik' uh noh' sis), which can result in fever, muscle soreness, and swelling. If not treated, it can even kill a person!

Viral Villains

Viruses are another kind of pathogen. They aren't really alive, but they do cause infections that can make you quite sick. They are not cells , like bacteria. They are more like very complicated chemicals. Because they are not alive, they can't reproduce to make more viruses. Instead, they invade a cell in your body and take control of the cell. They use the cell's capabilities to copy themselves over and over, until the cell is so worn out and full of viruses that it pops, spreading the virus – which then goes out and finds new cells to control. This continues on and on. No wonder you're so tired when you have a cold! Viruses cause lots of diseases, including colds, influenza (the flu), mumps, and AIDS. We'll discuss these more later in this lesson.

Viruses are often spread through the air when people

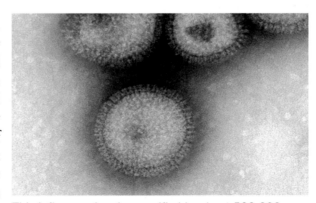

This influenza virus is magnified by about 500,000x so you can see it. It looks a bit like a cell, but it is not a cell. It is about one hundred times smaller than a bacterium.

cough or sneeze. They can also be spread when an infected person leaves viruses on the things he touches, like doorknobs and grocery carts. So during cold and flu season, wash your hands frequently, using soap and running water. Also, keep your hands away from your face. This will help keep you from picking up a virus and infecting your body with it. It will also help protect others from catching a virus you may have!

Cursed Cancer

You may remember that a cell reproduces by making copies of itself. This copy, of course, includes the DNA in its nucleus. But sometimes the DNA is not copied correctly. When this happens, the change in DNA is called a **mutation** (myoo' tay shun). If the part that is copied wrong causes the cell to start reproducing abnormally, a tumor is formed. Sometimes the tumor is just an annoying growth, but other times it results in cancer. Cancer is definitely a bad guy, but scientists are busy finding its cure. In fact, because of the progress made in cancer research, many, many people have survived its attack.

Dastardly Disease

A **disease** is something that upsets the normal workings of your body's systems. Although the word disease seems so frightening, it just means something isn't working perfectly in the body. In fact, the common cold is a disease. So, don't let the word disease scare you. Diseases can be short lived, which we call **acute** (uh kyoot') – even though they're never cute! They can also be long-lasting diseases, which we call **chronic** (kron' ik). Usually, both acute and chronic diseases can be cured. A disease can also be **malignant** (muh lig' nent), which means it can be life-threatening. But a disease can be **benign** (bih' nyne), which means it does not really harm you in any meaningful way. Diseases can be **contagious** (kun tay' jus), which means people can spread them to one another, but they can also be non-contagious, which means you can't get them from someone else. As you can see, diseases take on many different forms.

Some diseases are named after the body part they affect. Heart disease is one of these. Some diseases are named after the bacterium that causes the disease. For example, a **staph** (staf) infection, which happens when a cut gets infected, is caused by *Staphylococcus* (staf' uh loh kok' us) bacteria.

Infectious diseases are caused by parasites such as bacteria, fungi, worms, and viruses. Because there are so many infectious bad guys in the environment, most everyone catches at least one infectious disease a year. Fortunately, it's usually just a cold, which is a mild viral infection. But if you don't brush your teeth regularly, bacteria living in your mouth can cause cavities in your teeth, and cavities are a lot more trouble than a cold!

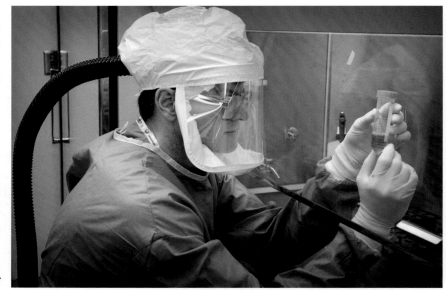

There is a virus in the tube this scientist is holding. He is wearing a lot of protective equipment because the virus causes a nasty infectious disease, and he doesn't want to be exposed to it.

Our Faithful Father

Now you know why we need to learn about the lymphatic system. It is your disease-fighting system that your faithful, heavenly Father built right into your body. God knew evil would come into the world and, thankfully, He made sure that our wonderfully designed bodies are equipped to deal effectively with most diseases on this earth. However, if you belong to Jesus, you'll one day receive a perfect body in heaven – a body that will never experience disease, wear out, or die. So, although there is bad stuff out there, one day we won't have to worry about any of it anymore. Heaven will be a much nicer place than earth. I hope I meet you there!

Before moving on, tell someone what you've learned so far.

The Lymphatic System

Let's find out what special mechanisms God put in place to keep you safe from diseases and pathogens like the ones we just discussed. Well, one thing God did was to fill your body with tiny masses called **lymph** (limf) **nodes**. They are found everywhere in your body. Think of a time when you were sick. Did your doctor put his fingers on either side of your throat to feel your neck? If so, he was feeling your lymph nodes. Did he look down your throat? He may have been looking at your tonsils, which work much like your lymph nodes. You've probably noticed that your tonsils swell a bit and become tender when you have a sore throat. You'll understand why in just a moment. Years ago, if tonsils swelled too often, people had them removed. As people become adults, their tonsils often shrink and become difficult to see. Do your parents still have their tonsils? Many adults do not.

The lymphatic system is made up of lymph nodes, **lymph vessels**, and organs like the tonsils and the **spleen**. Do you see that the diagram of the lymphatic system looks a bit like the diagram of the circulatory system? Well, that's because lymph vessels are usually found near blood vessels.

The lymphatic system is filled with fluid. Lots and lots of fluid passes in and out of the lymph nodes, through all those lymph vessels you see in the diagram, through the spleen, and everywhere in between. From where does this fluid come? Believe it or not, it comes from blood traveling through the circulatory system. Now you see why lymph vessels are located so close to blood vessels.

This is the way it works: As blood runs through the tiny capillaries, small amounts of clear fluid from the blood leak out into the tissues, keeping the tissues moist. This fluid, called tissue fluid, continually leaks out, so it must be collected. Otherwise, your tissues would become waterlogged! The tissue fluid drains back into the lymph vessels. As soon as the fluid enters the lymph vessels, it is called **lymph**. The lymph is filtered, cleaned up, and sent back to the circulatory system. Let's see how this happens and why it's so important.

On its way to the circulatory system, lymph fluid from the lymph vessels passes through one or more bean-shaped lymph nodes. Look at the diagram of a lymph node on the next page to see what one looks like. A lymph node is a bit like a filter. Do your parents have a water filter in the refrigerator or sink? If you've ever compared the taste of filtered water to regular tap water, you know the difference a filter can make. The filter removes impurities from your water so it tastes better. Well, that's what the lymph nodes do! Sometimes impurities enter your bloodstream. Your lymph nodes are designed to remove them from the lymph before it is returned to the bloodstream.

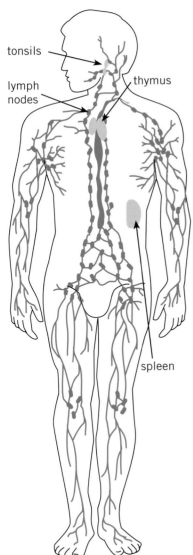

tonsils

lymph nodes

thymus

spleen

The lymphatic system looks a bit like the circulatory system.

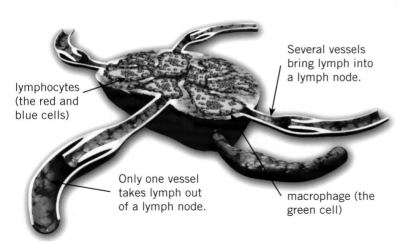

lymphocytes (the red and blue cells)

Several vessels bring lymph into a lymph node.

Only one vessel takes lymph out of a lymph node.

macrophage (the green cell)

Several lymph vessels carry lymph into a lymph node to clean it. Only one vessel carries the cleaned lymph away.

You have over five hundred of these little bean-shaped filters throughout your body. They work day and night to get rid of the impurities and make the lymph fresh and clean. These little filters are less than an inch wide, but they are very hard at work to help keep you healthy. Though you have them everywhere, God placed clusters of them in key locations. These clusters are found in your armpits, groin area, and neck. You can actually feel these clusters under your skin, especially if you are sick. You see, if you have some harmful germs running around your body, the lymph nodes nearest the germs get swollen and sore.

Inside the lymph nodes are invader detectors. They are white blood cells of a special sort, called **lymphocytes** (lim' foh sytz). They are on the lookout for any foreign invaders like bacteria, viruses, or fungi. They even detect cancer cells. If they detect even the smallest teeniest tiniest particle of an invader, they fly into action! Suddenly, they produce more and more lymphocytes, which are immediately sent out into the bloodstream on a search and destroy mission. This is much like an army called into duty when the smallest shred of a foreign invasion is discovered.

As the invader comes through the lymph into the lymph node, it is trapped there. Then, because the node begins making huge numbers of lymphocytes to search out other foreigners, the lymph node begins to swell. God designed lymph nodes with a stretchy elastic covering. This allows them to expand when they have to hold a good number of lymphocytes. When the battle is won, the lymph node eventually returns to its normal size.

Inside the lymph nodes are smaller nodules containing macrophages. You may remember these are white blood cells that eat up other cells. Macrophages engulf and destroy foreign material. As you can see, the lymph system is a terrific system designed to stop invaders dead in their tracks!

The Spleen

The spleen is located in the upper left abdomen and is about the size of a clenched fist. Do you recall another organ that's about that size? If you think back to the cardiovascular system, you might remember. If you guessed the heart, you're right! The heart and the spleen are about the same size. The spleen actually looks like a really large lymph node, and it happens to be the largest of the lymph organs. Like lymph nodes, the spleen is covered with stretchy connective tissue and is divided into smaller lobes. And guess what its job is? To filter impurities, just like a lymph node! But instead of filtering lymph, the spleen filters blood. Do you remember learning that red blood cells have a short life span? Well, when they die, they are consumed by phagocytes and taken to the spleen. The spleen removes millions of old or damaged red blood cells every second!

As in the lymph nodes, lymphocytes are produced in the spleen, and this production increases rapidly if invaders are recognized. The spleen also serves as a storage area for extra blood. Sometimes your body suddenly needs extra blood flow, like when you are involved in a physical activity, or if you sustain an injury. The extra blood needed is immediately squeezed from your spleen.

Did you know that it is possible to live without your spleen? Sometimes a person's spleen is removed because of an illness or an accident. There are even conditions that can cause a person's spleen to rupture. My coauthor (Dr. Ryan) once had a patient who was hit by a drunk driver while she was crossing the street. The driver had too much alcohol and could not react quickly enough to avoid her. This poor patient had to have her spleen removed because the surgeons could not stop it from bleeding. Within a few weeks, she recovered. Thankfully, God designed a backup plan for the removal of dead blood cells. When the spleen is gone, the liver removes them. However, without the spleen, the body is more prone to infections, so people without a spleen must regularly take antibiotics as a precaution.

Thymus

The thymus is in the chest, near the heart. This special organ provides the equivalent of a boot camp for lymphocytes. Many young lymphocytes are sent here for specialized training. They are taught special tactics and techniques for spotting and attacking foreign cells. These special forces are called **T lymphocytes** or **T cells**. In both cases, the "T" is for "thymus."

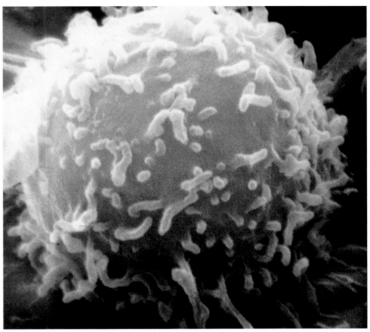

This T lymphocyte has been magnified by about 10,000x so you can see it.

Explain what you have learned about the lymphatic system so far.

Immunity

God has designed a wide range of defenses to protect your body. For example, like a protective wall around a city, the skin is an excellent barrier against the billions of bacteria found on its surface. These guys are just waiting for a break in the wall so they can stream in! Thankfully, the sebum produced by the oil glands is an antibacterial oil that continually destroys these lurking creeps. God even designed our tears to be an antibacterial moisturizer. In addition to these defenses, the mucus and cilia in the nose trap invaders and move them towards the throat, where they can be swallowed into the acid of the stomach to be destroyed. Finally, there are helpful bacteria in your intestines. Of course, they're only helpful as long as they stay in the intestines where they belong. If they get out, they are pathogens. As long as they stay where they belong, however, they are protective of their territory and will usually crowd out bad bacteria that find their way into the intestines.

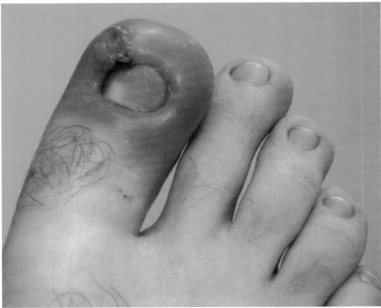

This person's ingrown toenail is infected, so the complement system has caused inflammation in that area.

All these defenses are **innate**, which means they are in your body all the time, and they respond to every threat in the same manner. The innate defenses are like moats and sturdy walls that keep foreigners from entering a castle and harming the people. These defenses make up your **innate immunity**. But God designed other systems just in case they can't keep the invaders out.

The second line of your innate immunity is called the **complement system**. This is composed mostly of proteins that are made in the liver. These proteins enter the bloodstream but are usually inactive. If they come into contact with foreign bacteria, however, they become activated. When activated, they cluster together and poke holes in the cell membranes of the bacteria. The fluid surrounding the bacteria rushes in, and the bacteria explode. It's dinner time for the macrophages! They eat up the pieces of the exploded bacteria!

When these proteins become activated, they also send out chemicals that attract white blood cells. These white blood cells cause **inflammation** (in' fluh may' shun), which swells the tissue. If you've ever had a splinter surrounded by red, swollen, and tender skin, you've experienced firsthand how annoying inflammation can be. But inflammation is an extremely important part of the immune response in part because it is so annoying! It directs your attention to an area of your body that needs urgent care, so that you can give your immune response a helping hand – in this case, by removing the splinter.

Inflammation does many things besides causing tenderness and swelling. One is to increase the local blood flow to the area of injury or infection. When tissue is injured, special kinds of chemicals called **histamines** (his' tuh meenz) are released and sent to the injured tissue. These histamines cause the capillaries to become wider, which allows more blood to flow through. The increased blood flow causes the area to become warmer and redder. At the same time, the capillary walls temporarily become more porous, causing more fluid leaks from the vessels into the surrounding tissues. This causes tissue swelling, and at the same time, it makes it easier for the arriving white blood cells to slip out of the capillary and into the surrounding tissue.

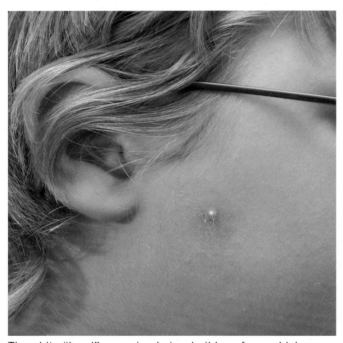

The white "head" on a pimple is a buildup of pus which comes from your body destroying bacteria that have been trapped in the pimple.

Let's imagine again that you have a splinter that has been difficult to get out. If the splinter carried a large number of bacteria (and they usually do), pus will form. This pus is made up of damaged tissue, living and dead bacteria, and dead white blood cells. Large collections of pus will eventually drain through the skin; smaller collections will be carried away by macrophages. If the amount of infected tissue is large, your body will heat up, resulting in a fever. Fevers are designed to help destroy bacteria or viruses in your body. That's why the doctor asks if you're running a fever when you are sick. The higher body temperature caused by the fever increases chemical activity, and many invaders find it more difficult to survive in a warmer environment. Your own white blood cells and inflammatory proteins can tell the brain to increase your body's temperature; certain foreign invaders can do the same thing. This increase in temperature makes you tired, so you often need to rest until the battle is over.

Do you remember that red blood cells, white blood cells, and platelets are all formed in the red bone marrow? You might also remember that I mentioned stem cells in an earlier lesson. Do you remember that stems cells are cells that can become many different kinds of cells? Some of the stem cells in your bone marrow will develop into white blood cells that do the same activities all the time. Because they do the same activities all the time, they are part of your innate immunity.

Some of these white blood cells will eat foreign material to destroy it. Some will release chemicals that help in the healing process. The histamines we mentioned earlier, for example, come from such cells. Other white blood cells release chemicals that are poisonous to the invaders (but not to you). There are even white blood cells that will recognize when one of your cells has been infected, and they will kill that cell in an effort to keep the infection from spreading. All these white blood cells are very effective at what they do, so they continue to do the same thing all the time, no matter the nature of the foreign invaders.

Special Agents B and T

Some stem cells become white blood cells that make up your innate immunity, but some stem cells develop into another group of warriors. These are like special agent cells that wage a different kind of war. They have a more flexible and adaptive response that can deal with lots of different kinds of threats in individual ways. It's as if these cells can actually think and respond according to the particular situation. The special agents are **B lymphocytes** and **T lymphocytes**. B lymphocytes (often called B cells), develop in the bone marrow and then move about the body in the bloodstream or stay in reserve in the lymph nodes and spleen. The T lymphocytes, or T cells, head to the thymus to finish developing. The "T" stands for thymus, while the "B" stands for bone. So, they are named after where they were "trained" to fight.

The specialized B and T cells have the ability to recognize cells that are "self" and cells that are "non-self." The "self" cells belong in the body. The "non-self" cells are invaders (or cancer cells) that do not belong in the body. Even more, these special agent cells have a memory! They can recognize and remember specific invaders, so that the next time that particular germ tries to invade again, they react with a very rapid, specific, and aggressive response. These are the forces responsible for making sure you don't get chicken pox a second time. This is called **adaptive immunity**. Both B cells and some special T cells are part of this sophisticated system.

Antibodies and Antigens

To understand how adaptive immunity works, you need to know about **antibodies** (an' tih bod' eez) and **antigens** (an' tih junz). You may remember that we used the word "antigen" when we talked about blood types in Lesson 7. Antigens are markers, like the flags on a ship that proclaim to all the world the country of the ship. The cells in your body have multiple antigens that mark the body's cells as "self." The combinations of markers are as unique as your fingerprints. Some of the markers are much larger than others. Any antigen that is "non-self" belongs to an invader, and the cell flying that flag is soon discovered and removed by your immune system.

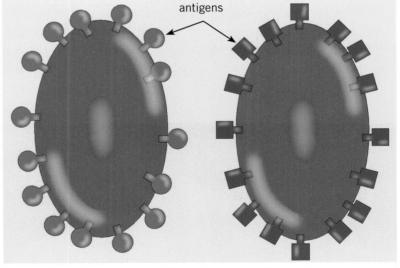

Because these two cells have different types of antigens, they come from different organisms.

What's really interesting is how the cells communicate with each other about foreigners. This is how it works: When macrophages eat pathogens, they actually save some of the flags (antigens) from the pathogen. They then take those flags to a B cell or a special T cell called a helper T cell and hold them high for the B cell or helper T cell to see. The cell recognizes the flag as being from a foreign, "non-self" invader, and begins to plan for battle.

B cells begin making special proteins called **antibodies** that will grab the specific antigens. They tailor the antibody to fit the invader exactly. In other words, the antibody will grab on to the antigen of the invader, but it won't grab on to any other antigen. After creating the antibody, the B cell copies itself, summons other B cells, and continues to produce the particular antibodies to fight against the specific antigen it has seen. Many

B cells will be activated at the same time if there has been a big invasion.

Once the specifically-created antibodies are made, they go out and lock onto the foreign antigens, flagging the foreign cell for destruction. These "non-self" cells are then destroyed by the complement system, by macrophages, or by special T cells called **killer T cells**. Anyone who has ever been sick knows this whole process can take days, weeks, and sometimes years if the invader is particularly wily or if the immune system has been weakened. And sometimes, because we live in a sinful world under the control of the evil one, the invaders win. But thankfully, this doesn't happen very often. Although we are bombarded, sometimes daily, with foreign pathogens, our God-given special agent cells do all they can to keep us safe.

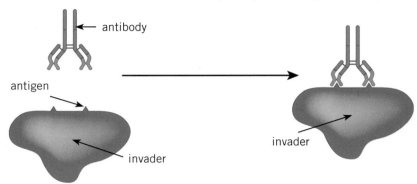

Once an antibody has been made for a specific invader cell, the antibody will latch onto the invader's antigen, which marks the cell as a "kill target" for other cells.

Once the battle is over, those B cells and T cells that weren't destroyed will demobilize, but they continue patrolling. Some of them will retain the memory of the invader's flags and be ready if another invasion occurs. This entire response system is very closely regulated so that invaders, but not "self" cells, are attacked.

Do you remember that viruses invade your "self" cells? They sneak inside and secretly take control, forcing the cell to make more and more of the virus, all the while wearing the cell out with such rapid reproduction. In addition, cancer causes cells to copy themselves way too much, making a tumor. Fortunately, your immune system has specialized training to deal with these evil sneaks. While in the thymus, T cells are trained to recognize the body's cells as "self" and to not attack those cells. But when a virus takes over a cell or a cell is cancerous, it cannot hide completely. Antigens of the virus or cancer cell are visible in the cell membrane, and the T cells can see them. They attack these cells and destroy them.

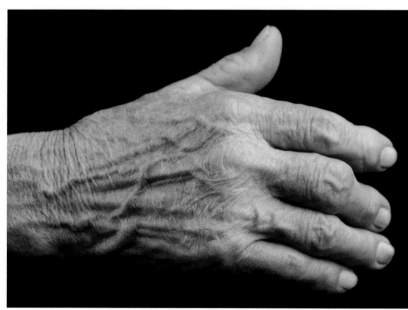

This hand belongs to a man who has rheumatoid arthritis, an autoimmune disease in which the body's immune system attacks his own tissues in his joints.

Antibody Antics

Sometimes our antibodies overreact. This is part of living in a fallen world. Some of these antibody antics cause **allergies** and **autoimmune** (aw' toh ih myoon') **diseases**. Autoimmune diseases are those in which the body forms antibodies against its own tissue. Doctors treat these diseases by using medications that make the immune system less active. Unfortunately, this also makes infections more likely. Hopefully, better treatments for autoimmune diseases will be discovered soon. Allergies occur when your B cells begin to make antibodies to fight foreign substances that are not harmful viruses, like pollen and pet dander. When these substances get inside your body, your immune system overreacts – treating the substances like foreign invaders. This causes you to feel like you have a cold, because your body is responding as if you do! Allergy sufferers are often given antihistamines, which combat the inflammatory actions of the histamines released by the person's innate immune system.

HIV (human immunodeficiency virus) is a virus that actually infects white blood cells, especially helper T cells. It destroys these defense cells so that a person with HIV is unable to fight against even the most basic illness. If HIV is not defeated, it can cause the syndrome known as AIDS (acquired immunodeficiency syndrome). People with AIDS can die from illnesses that are not normally deadly, because the virus has destroyed so many of their immune-system cells. Many scientists are looking for a cure for AIDS. Hopefully they will find one soon.

Immunity Modes

During pregnancy, many of the mother's antibodies are passed to the baby and get into the baby's bloodstream. This is called **passive immunity**, because the baby acquired it without any action on the part of the baby's immune system. Babies also receive passive immunity through their mother's milk. Incredibly, if the baby has a cold that the mother hasn't yet caught, her milk will develop specific antibodies to fight the baby's cold – even though she doesn't have it! This can continue until the baby is around six months old.

Another form of passive immunization happens through antibody transfer. Antibodies are filtered from the blood of a donor who has been exposed to a particular infection. These antibodies are then transferred to another person who may have been, or possibly will be, exposed to that particular infection. This transfer of antibodies will give the recipient protection from that particular pathogen for a while. However, since the recipient's immune system didn't actually experience the pathogen, his immune system won't have a memory of the proper immune response, and he could get the pathogen later on in life. This type of passive immunization is only done when the infection could cause significant illness, and the person has no immunity to the disease. For example, if you are bitten by a bat or a rabid dog, you'll be vaccinated against rabies. But the vaccination takes weeks to stimulate your immune system, and rabies is generally fatal well before that. So antibodies against rabies are injected around the bite site to stop the infection.

Acquired Immunity

Acquired immunity occurs when your immune system is exposed to, and responds to, a specific threat. Once you've had chicken pox, for example, you don't get it again. Your memory B cells and T cells recall the invasion, and if any chicken pox viruses come your way, they are immediately destroyed. In the case of rabies, the vaccine will produce acquired immunity, while the shot of rabies *antibodies* produces passive immunity. Some viruses, like influenza (the flu) and many common colds, change their flags (antigens) frequently. This changing of the flag is usually caused by a mutation. Because of this, it's not possible for you to develop lifelong immunity against the flu and colds. Flu viruses mutate every year, as if flying different flags that fool your body.

Because the common cold virus mutates so often, your body cannot build up an acquired immunity against a cold.

Vaccinations

You've probably heard of vaccinations, which are shots designed to prevent infections. Basically, they contain killed or weakened pathogens. Some actually contain chemicals that simply have the same flags that the pathogen has. They serve to stimulate the immune system in an artificial way. The body isn't actually fully infected, but it develops the antibodies to fight the disease anyway. The immunity generated by vaccinations is called **active artificial immunity**. Because the immune system has memory, this active immunization can last for a long time. However, boosters are sometimes required in order to ensure the immunity continues over many years.

Children used to die regularly from a disease called smallpox, which is similar to chicken pox. Once a proper vaccination was developed, people stopped getting smallpox. Because of the vaccination, this disease is no longer a threat. In fact, the only place you can now find the virus that causes smallpox is in a laboratory. The smallpox vaccination simply kept the virus from infecting people, so the virus could no longer copy itself. As a result, it eventually decayed away. Since the virus no longer exists naturally, smallpox vaccines are no longer given.

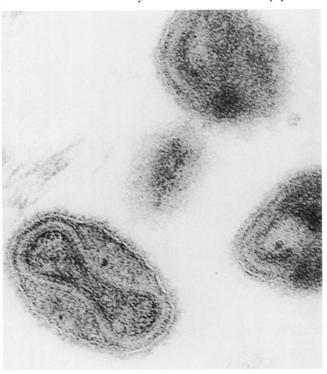

This magnified image (colors artificially added) shows the virus that causes smallpox. Because of effective worldwide vaccination, this virus now exists only in laboratories.

The kinds of diseases that children are usually vaccinated against are tetanus, measles, polio, whooping cough, rubella, and mumps. Tetanus, also called lockjaw, is a terrible disease that is easy to get when a deep wound is infected. The germs that cause this infection are found in the soil of every country. Tetanus causes painful muscle contractions and death. Vaccination is highly effective at preventing this disease. Measles doesn't usually cause death, but it can cause permanent brain damage, and it is definitely not any fun to have. Polio can result in flu-like symptoms, and it can leave you crippled for life. Whooping cough (also called pertussis) is common in adults, but if infants and children get it, they cough for months and can suffer brain damage from this disease.

In countries that vaccinate, diseases like tetanus, polio, measles, and severe whooping cough are very rare. Because we rarely see these diseases anymore, some people choose not to vaccinate their children against them. Often, the reason they choose not to vaccinate is because vaccinations might carry a risk of certain side effects. However, today choosing not to vaccinate also carries risk. This risk is real because of air travel from non-vaccinating countries, which might result in the diseases returning. For example, on May 15, 2005, an unvaccinated 17-year-old girl returned to the U.S. after a mission trip to Romania. What she did not know was that she was carrying the measles virus back home, too. She attended a large gathering of fellow church members the day after she got home, and she spread the disease, mostly to the unvaccinated people in attendance. There were a total of 500 people at the gathering, and 34 people ended up coming down with the measles. Three of the 34 infected people were hospitalized, one of whom required six days of ventilator support in order to survive. Thankfully, there were no deaths.

Amazing Antibiotics

As you've learned with vaccinations, we sometimes need a little help dealing with the diseases that our bodies are having trouble fighting. Thankfully, God has given us the ability to discover many medicines that help our bodies fight the battles they will encounter. One of these special discoveries was **antibiotics** (an' tye bye ot' iks). The word comes from two Greek word roots: "anti" means against, and "bios" means life. These chemicals kill living things. When made into medicine, they are very effective at killing unwanted living organisms that invade our bodies. In nature, antibiotics are released from fungi or bacteria in order to kill the surrounding competition.

This is a magnified image of *Penicillium*, the mold that produces the antibiotic known as penicillin.

Penicillin (pen' ih sil' un)) was the first antibiotic ever used. It comes from a mold (a type of fungus) called *Penicillium* (pen' ih sil ee' uhm). The ability of the *Penicillium* mold to kill bacteria was discovered in 1928 by Alexander Fleming, a British scientist. He was growing bacteria for a study, when he realized that mold growing in his lab was killing the bacteria. Then it dawned on him – something kills bacteria! And penicillin was discovered. It wasn't easy to make.

Later, in 1932, another type of antibiotic was discovered: sulpha powder. Sulpha powder was much easier to make than penicillin. During WWII, when British and American scientists began searching for a way to keep soldiers from dying after being wounded, they sprinkled sulpha powder on the wounds. Over the years, many other antibiotics have been discovered.

So now you know all about the amazing defense systems that God designed to keep you healthy and strong. But remember, the better you take care of your body, the better those defense systems work!

What Do You Remember?

What is a pathogen? What is an infectious disease? Name one kind of disease. How do lymph nodes protect us from diseases? Where can stem cells that develop into disease-fighting cells be found? Name some of the systems that help keep us well as our first line of defense. What name is given to that part of our immunity? What does a fever do? What do we call the immunity we get from cells that analyze attackers and remember them in case they come back? How are antibodies formed in the body? How do vaccines work? How was penicillin discovered?

Notebooking Activity

Write down all that you wish to remember about your body's defenses. After that, create some pages for your notebook that detail information about pathogens, the lymphatic system, and the different mechanisms of immunity. Make illustrations to go along with your descriptions.

Experiment
Testing for Bacteria and Fungi

We have mentioned many times in this course that bacteria are everywhere. Thankfully, God designed your defense systems so well that the bacteria usually don't harm you at all. So, you needn't be afraid. In this experiment, you will use a prepared kit with a solution called agar to find out which items in your house have the most bacteria. You will also do an experiment to discover the best way to clean bacteria off of surfaces.

You will need:

- A bacteria testing kit with agar (ah' gar) and Petri dishes (The course website I told you about in the introduction to the book has several links to websites that sell them for under $20.)
- Q-tips
- Tape

Think about places around your home like doorknobs, kitchen counters, bathroom faucet handles, etc. Which do you think have the most bacteria and fungi? Choose as many places as you have Petri dishes in your kit. Those will be the places you test for bacteria. Make sure that one of the places you choose is the inside of your mouth. Make a hypothesis about which surfaces will have the most bacteria and fungi and then write it down on a Scientific Speculation Sheet. Now let's do the experiment to see if your hypothesis was correct.

1. Follow the package's directions for preparing the agar.
2. Once it is prepared, pour the agar into Petri dishes, cover the dishes, and allow the agar to harden overnight.
3. The next day, use a Q-tip swab to sample one of the locations you chose to test. Just rub the swab on the thing you want to test so the Q-tip picks up whatever is there.
4. Rub the swab across the hardened agar in a Petri dish. This will transfer what was on the surface to the agar, and the agar will feed any bacteria or fungi, allowing them to grow.
5. Cover the Petri dish and tape the cover on the dish so it does not come off.
6. Label the cover with the name of the place you swabbed with the Q-tip.
7. Repeat this for every location you want to test, using a new Q-tip each time.
8. Wait a few days to see bacteria and fungi grow in the Petri dishes. **BE VERY CAREFUL WHEN DEALING WITH BACTERIA! WASH YOUR HANDS THOROUGHLY AFTER TOUCHING THE PETRI DISHES, AND DISCARD THE Q-TIPS.**

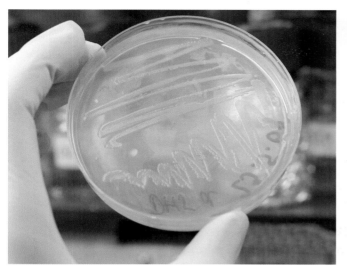

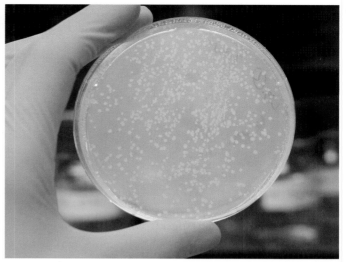

When you rub the Q-tip over the agar, it will probably leave "tracks" behind. That's fine. You need to get the bacteria and fungi off the Q-tip and onto the agar.

Individual bacteria are too small to see without a microscope. However, a large number of bacteria will show up as dots, lines, or discolorations on your Petri dish. Fungi will also show up this way. The more dots, lines, or discolorations, the more bacteria and fungi.

9. Which location ended up producing the most growth on the Petri dishes? That location probably had the most bacteria and fungi on it.

10. Kill the bacteria and fungi in the dishes by washing them with bleach. Although most of the bacteria and fungi growing in your house shouldn't make you sick, it's better to be safe than sorry.

11. If you wish, you can extend this kind of experiment to see what kills bacteria and fungi. For example, you can swab an area that you know has a lot of bacteria on it, and then you can put it on agar that also has something antibiotic on it (like antibiotic soap). If you compare the growth on that agar to the growth on agar with no antibiotic on it, you can get an idea of how well the antibiotic actually reduces bacterial and fungal growth.

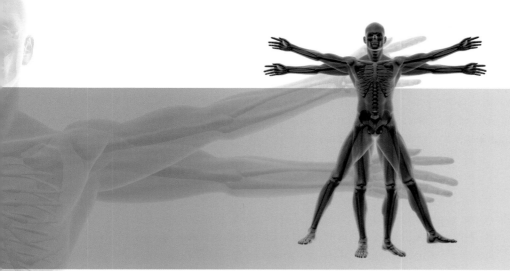

GROWTH AND DEVELOPMENT

Before you were born, God knew you and loved you. He watched over you as you grew from one single cell into a singing, laughing, thinking, worshipping member of His family. And God is still watching over you, even now. He's thinking about you all the time, and His thoughts toward you are precious!

All parents wonder what their children will look like, be like, and do in the course of their lives. Children generally wonder about these things as well. Have you ever wondered why you look the way you do? Do you ever think about growing up? Do you wonder when you will begin to change from a child into an adult? Have you considered what God's plan is for your life?

O LORD, You have searched me and known me...
And are intimately acquainted with all my ways.
Even before there is a word on my tongue,
Behold, O LORD, You know it all.
You have enclosed me behind and before,
And laid Your hand upon me.
For You formed my inward parts;
You wove me in my mother's womb.
I will give thanks to You,
for I am fearfully and wonderfully made;
Wonderful are Your works,
And my soul knows it very well.

Psalm 139:1, 3-5, 13-14 reminds us that we are made by God.

In this final lesson, we'll discuss these things. We'll talk about how you grow, how you develop, and how you got the genes that affect the color of your hair and eyes. We'll also talk about what makes people different from animals and the truth about how God made man. You'll find out that not all people believe God made man, but instead believe man came from animals. You'll learn some of the reasons why they believe this, and how those reasons have been shown to be false. Most importantly, you'll discover that God has a special purpose for creating you and a special plan for your life! Let's begin the last leg of your journey into human anatomy and physiology.

Dividing Cells

Do you remember learning that cells come together to form tissues, tissues come together to form organs, and organs form systems? In the previous lessons, you learned how many of those systems work. You've really learned a lot this year, haven't you? Well, let's not stop there! There's more to learn! Have you ever wondered how your body can grow these tissues, organs, and systems? Well, it all begins with a single cell. But how does that single cell become an entire human being?

God created your cells to divide. Let me explain. You started out as one cell. That original cell pinched itself in half and became two cells. Those two cells, instead of remaining two smaller cells, actually grew to the same size as the original cell. So by dividing, your one cell actually became two! Your cell doubled! This doubling occurred again and again: one cell made two cells, then two cells made four cells, four cells made eight cells, eight made sixteen, and on and on until you had trillions of cells that make up you – and that's how your body grew! The miraculous thing is that the cells not only grow, but they change into different types of cells, like nerve cells, bone cells, and skin cells. After forming into different cells, they move to the proper areas to become organs. And it's a good thing too, because you wouldn't want your brain to be in your knee or your heart under your arm! Thankfully, God made you so that everything works out just right almost every time.

God designed every cell to do exactly what it's supposed to do, even to know what size it should be. This ensures you don't turn out to be the size of a flea or the size of a house! Your body knows when to grow rapidly (during infancy), when to slow down (age three until about age 10), when to grow rapidly again (from about 10 to about 14), then when to stop growing (upward at least! – when you reach your adult height). It would be awfully strange if you grew to be bigger than your parents at two years old. However, when you are more mature and ready for more responsibility, your cells begin dividing more rapidly again so you can become an adult.

Did you know that growth actually happens in spurts? You grow from one microscopic cell to seven or eight pounds before you are even born. After you are born, you grow very fast until you are about three years old. Then, you slow down. Between the ages of 10 to 14 years old, you start growing like a weed. These spurts continue until you're about 16 to 18 years old, perhaps as late as 20 years old if you are a boy. At that time, you reach the full size you were designed to be.

Different parts of this baby's body will grow at different speeds. His head, for example, will grow slowly compared to the rest of his body.

What's even more interesting is that different parts of your body grow at different speeds. Your shape changes as you grow up. For example, as a baby, your head was about 25% of your total size. When you grow to be an adult, your head will be only about 10% of your total size. Now of course, your head doesn't shrink as you get older; it just doesn't grow as fast as the rest of you. Look at the picture on the preceding page. By the time the infant is an adult, his body will be more than three times as tall as it is in the picture. However, his head is already about half the size of his father's head. That tells you his head will not grow as fast as the rest of his body.

Try This!

Gather some photos of when you were a baby, a toddler, and a small child. Compare them to a recent photo of yourself. How have you changed? How have your body proportions changed?

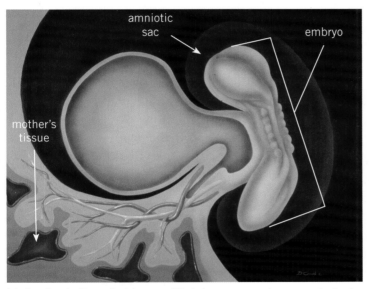
It's hard to believe, but in your 21st day as an embryo, you looked something like this and were just over one-tenth of an inch long!

Development in the Womb

Before you were born, you did most of your developing in your mother's **womb**. At the start, all your cells were the same. Do you remember stem cells? As a newly-forming baby, you were made up mostly of stem cells. Can you guess why? It's because a stem cell can become any kind of cell. That first stem cell copied itself over and over. All those copies formed into a hollow ball. After about two weeks, the cells began to change into specialized cells and leave the hollow ball, moving to where they would form into your head, shoulders, knees, toes, and everything in between. You were called an **embryo** (em' bree oh) at this point. It's obvious from all this cell activity that – even though you didn't look like yourself – you were very much alive. You lived in a fluid-filled bag called an **amniotic** (am' nee ot' ik) **sac**, which was attached to your mother.

On day 25 of your life as an embryo, your heart started beating! All the cardiac muscle, blood vessels, nerves, fat tissue, cartilage, and other tissues were perfectly woven together in your little body in order to pump your blood, which was being formed in a "sac" that was attached to you. Yes! You even had blood when you were less than an inch long! Your nerves had been developing for a long time, but by day 28, the basic layout of your central nervous system was complete. You even were developing the first stages of your gastrointestinal tract! By day 32,

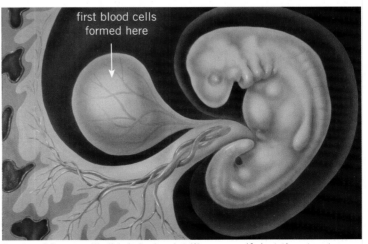
By day 28 you still didn't look a lot like yourself, but the structures that eventually formed your head and eyes were recognizable. At this point, you were a whopping one-fifth of an inch long!

your liver began to form. Think about that! After just 32 days from when you were a *single cell*, you had all the makings of your circulatory system, central nervous system, digestive system, and liver!

In the second month of your development, all those organs that were formed started growing and becoming stronger. Also, you started taking the shape that you would have when you were born. Your limbs (arms and legs) started to develop, as did your fingers and toes. During this time, your first bone cells also started to form. Do you remember what we call the cells that lay down new bone? If you said "osteoblasts," you were right! Also, your inner ear started to form. Remember, that's what allows you to hear and to have a sense of balance. By the end of the second month, you were a bit more than an inch long, and you were called a **fetus**.

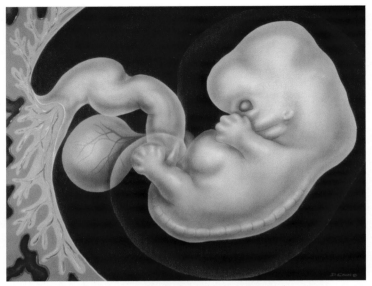

Midway through your second month as an embryo, you started developing the shape that you would have when you were born, and you were about half an inch long.

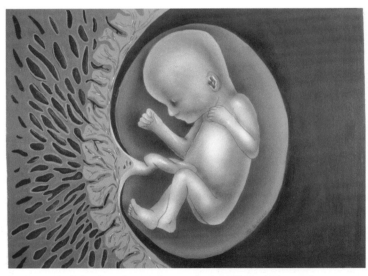

At the end of the fourth month, you looked like a baby and were about five inches long.

Three months after being a single cell, your head grew to be about half of your total body length. Your legs and arms developed enough so that you could move them, exploring the small world inside your mother. By the end of the third month, a good doctor could take an ultrasound image of you (an image made by using sound waves that pass through your mother's womb) and determine whether you were a boy or a girl. Your teeth also began to form.

In your fourth month, you actually started growing fine hair on your head! Your muscles started developing with gusto, and your bones started hardening. Because your muscles were developing a lot, you started moving a lot more. Your mother probably didn't feel you moving yet, however, because you were still pretty small.

In your fifth month, the hairs you started forming covered your entire body. Your eyelids and even eyebrows developed, and you actually started to hear things. Your sebaceous glands (remember, they make oil for your skin) and some of your endocrine glands (remember, they make hormones) started working. Your spinal cord started to develop a myelin sheath so that messages could run through it more quickly. Even though you started moving in your mother's womb before this, by the end of your fifth month, your mother probably started to feel your movements.

Through all this time, your skin was rather transparent, but in your sixth month, it became red and wrinkled. In addition, your bone marrow was developed enough to start making red blood cells so they no longer had to be made in the "sac" that was attached to you, and your alveoli (remember, the air sacs in your lungs) started to form. In your seventh month lots of nervous system development occurred. As a result, you were able to open your eyelids. In your eighth month, your eyes could actually react to light. Also, your bones were all formed, but they were still soft and pliable. However, during that month, your body started to store

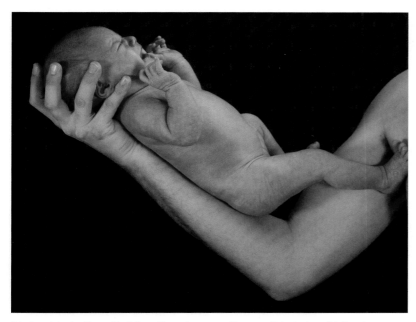

It's hard to believe that less than ten months ago this baby girl (on her mother's arm) was a single, microscopic cell!

the calcium it needed to begin making them harder. In your ninth month, the "finishing touches" like fingernails that extended beyond the tips of your fingers were formed, and around the end of your ninth month, you were born!

Development Outside the Womb

You did a lot of developing in your mother's womb, but it didn't end when you were born, did it? You started developing right away as an infant, and you are still developing to this day. When you were about three months old, you could probably hold up your head and look about. You could babble a bit and swipe at objects within reach. At seven months old, you could sit, use your hands like a rake to grab objects, and respond to your own name. By the time you were one year old, you were probably crawling everywhere and beginning to take a few steps. You could probably also say a few words, like "mama" and "dada." Of course, development is different for every child. My daughter could speak in sentences by the time she was a year old, but she couldn't walk until she was almost 15 months old. Much of your development is related to the myelin sheaths that surround your neuron axons – the more myelin you have, the faster your nerves can transmit information, and the better you can make your body do what you want it to do. Though you haven't gained many neurons since you were born, the connections between the neurons you have continue to form all your life. The more you read and learn and experience, the more connections you'll have, and the smarter you'll be!

By four or five years of age you could do almost anything! You could climb trees, learn to swim, and kick a ball. Between the ages of five and seven, you were able to understand complex information, like science! The amazing brain that allows all this learning began to develop about 21 days after the first cell appeared. Your brain continued to grow until you were about four years old.

As you get older, certain changes in your body and mind are specific to whether you are a boy or a girl. The major changes in this regard occur during **puberty** (pyoo' bur tee), and they usually begin between the ages of 9 and 13 for girls and between 10 and 14 for boys. During puberty, children grow much taller, and the differences between males and females become more obvious. In short, you are becoming a man or a woman so that eventually you can marry and have children of your own.

Of course, as you develop, you also grow. The word "growth" usually just refers to getting bigger, while the word "development" refers to how you change (both mentally and physically) as you mature into an adult. One of the best measures of growth, of course, is how tall you will be. This depends on the DNA that your parents gave you (we call that "**genetics**") as well as the food you eat, the way you are raised, how much exercise you get, and so forth (we call that "**environment**"). If both of your natural parents are tall and you eat nutritious food and exercise, it is likely that you will be tall as well. But if both parents are short, your chances of being short are pretty good.

Try This!

There are many formulas that doctors use to determine how tall a child will grow to be. The most accurate predictors I have seen can be found on the Internet, using the mother's height and the father's height, as well as the child's current height and age. So, if you want to find out how tall you might be when you grow up, go to the course website I told you about in the introduction of this book, and you will find links to some different height calculators.

Genetics

You may wonder why you would need the heights of both your mother and father to predict your own height. Well, it's because of genetics. You see, long ago people noticed that children looked like both of their parents but were not exact copies of either one. A female child is not an exact copy of her mother, nor is a boy an exact copy of his father. Yet children often resemble one parent in some traits, and the other parent in other traits. A trait is a recognizable physical characteristic, like eye color, skin color, or hair texture.

My husband and I were looking through old family photos that belonged to his grandmother. I noticed that even his great, great grandparents had his face, nose, and jaw structure. These traits were passed to him from his ancestors! Traits are determined by genes, which are found in DNA. Do you remember learning about genes and DNA in the first lesson, when you studied cells? Well, we are about to study DNA a little more seriously now.

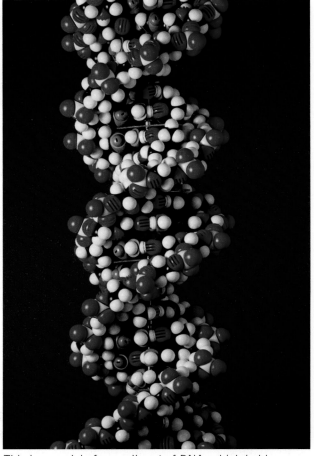

This is a model of a small part of DNA, which holds your genes.

Try This!

Before we jump into DNA, let's study your traits. You can easily find out which traits you share with your mother and which you share with your father. If you are adopted, you can still see if you have any traits that are similar to your adoptive parents. Which parent has the same color hair as yours? Which parent has the same color eyes? Do your eyebrows match those of either parent? Can you roll your tongue? Can one of your parents? Do you have freckles like mom or dad? Is your second toe longer than your first toe or shorter? Which parent has your toes? It's interesting to notice all these features, isn't it?

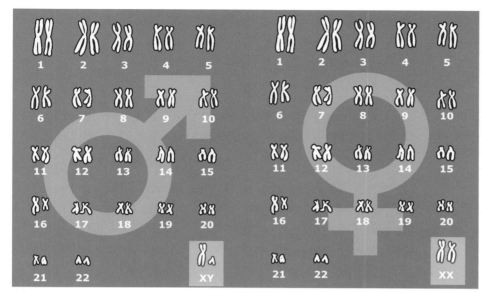

People have their DNA split into 23 pairs of chromosomes. 22 of the pairs look the same for men and women. The only pair that looks different (highlighted) determines whether or not you are a boy (XY) or a girl (XX).

Chromosome Commotion

Genetics is used in the study of **heredity** (huh red' ih tee), which tries to understand how traits are passed on from parents to their children. Do you remember that genes are contained in your DNA? Well, it turns out that your DNA is actually split into smaller units called **chromosomes** (kroh' muh sohmz). In fact, you (and people in general) have your DNA split into 46 chromosomes. Plants and animals have different numbers of chromosomes. Horses, for example, have 64 chromosomes, cats have 38, and one species of fern has 1,260 chromosomes! Each of a person's 46 chromosomes has thousands of genes on it. These 46 chromosomes are matched up in pairs. So, you have 23 pairs of chromosomes.

Every cell in your body contains DNA at some point in its life, which means at some point, every cell has chromosomes. If a cell has DNA, it has all the information necessary to be any cell in the body. However, in each cell, only a portion of this information is used – only the part that's important to that particular cell. When a stem cell becomes a skin cell, for example, the skin cell automatically knows which DNA information to use (the information necessary for a skin cell) and which to ignore (any information not related to being a skin cell). It's kind of like having a full library, but only reading the books you're really interested in reading.

Do you remember where the DNA is stored inside a cell? It's stored in the nucleus. Most cells in your body, no matter how microscopic they are, have about 9 feet of DNA tightly coiled inside the nucleus. The coil is made up of two strands that are twisted into a double helix (think of a ladder that twists as it goes up) – so that it looks like the model on page 236. Because an adult has trillions of cells, if all the DNA in an adult body were placed end to end, it could *stretch to the sun and back 70 times!* It's truly amazing that this enormous amount of information can be stored in tiny cells. What's even more amazing is the fact that that this DNA can be duplicated accurately millions of times a day so that cells can make more cells!

Merry Mitosis

Each time a cell divides to make a new cell, all the DNA must be transferred to both of the new cells. The way in which this happens is remarkable. The DNA untwists itself, section by section, and a new copy is made of each strand. Just before the cell divides, the cell has twice as many chromosomes as it started with, because all the DNA has been copied. The cell then divides, forming two new cells, each with a copy of the DNA. So, all the information from the first cell is passed on

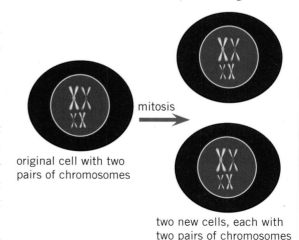

mitosis

original cell with two pairs of chromosomes

two new cells, each with two pairs of chromosomes

When a cell divides by mitosis, it makes two new cells with exactly the same DNA as it had.

to both of the new cells. This process is called **mitosis** (my toh' sis). With mitosis, all the genes are passed on to the new cells. A cell that has four chromosomes, then, will make two new cells, each with four chromosomes.

Magnificent Meiosis

So how do children resemble a mixture of both parents if the divided cells are exact copies of the original cell? Well, God designed a system in which the genes from the mother and father mix together. This is how you can end up with your father's big nose, your mother's green eyes, your grandmother's curly hair, and your great grandfather's ability to curl his tongue!

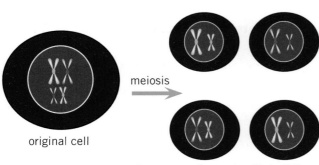

original cell

four new cells, each with half the chromosomes as the original cell

In meiosis, each new cell gets only one chromosome from each pair. If a cell starts out with two pairs of chromosomes (for a total of four), the new cells will each have only two chromosomes (one from each pair).

This occurs through a special cell division process called **meiosis** (my oh' sis). This division only happens when the cells that come together to make a baby are being formed. Meiosis starts like mitosis, with the copying of the DNA within a cell. But during meiosis, cells divide more than once; they divide twice. So, even though only two exact copies of the DNA were made, they have to be split up among four cells. As a result, the final cells end up with half the number of chromosomes as the original. If the cell started with four chromosomes, all four of the cells made in meiosis have only two chromosomes. How does the cell do this? It gives one chromosome from each pair to each cell. So, a cell that has two pairs of chromosomes will make four cells that each has only one chromosome from each pair. As a result, the cells will have only two chromosomes. In people, the cell starts with 46 chromosomes that are grouped into 23 pairs. After meiosis, each cell has only one chromosome from each pair, so each cell has only 23 chromosomes. These cells, with half the number of chromosomes that normal cells have, are called **gametes** (gam' eetz), or reproductive cells.

So…your mother has gametes, and your father has gametes. To make you, one gamete from your father combined with a gamete from your mother. Your father's gamete had 23 chromosomes – one from each of his chromosome pairs. Your mother's gamete had 23 chromosomes – one from each of her chromosome pairs. When they combined, they produced a cell with 46 chromosomes, and that cell developed into you! The two gametes literally become one, each providing half of the chromosomes you needed. So when did you become alive? It was when your parents' gametes merged to form a single cell that contained all the DNA that makes you what you are today. Even though you were only a single cell at that point, the rest of what needed to happen so you could be born was just development, and everything you needed for that development (except for food and oxygen – those were provided by your mother) was in that single cell!

two gametes come together to form a single cell

gamete from one parent

gamete from the other parent

new life!

When two gametes combine, the result is a new individual with the right number of chromosomes.

So you have 23 chromosomes from your mother and 23 chromosomes from your father, and together those 46 chromosomes hold all your genes. Where did those chromosomes come from? Well, your mother got 23 from her mother and 23 from her father, and your father got 23 from his mother and 23 from his father. Of course, your parents' parents got 23 chromosomes from each of their parents, and so on. Do you see that the genes you have were passed down through your entire family to end up in you? You have genes from your great, great, great, great grandfather! Isn't that amazing?

Now what's amazing is that each time meiosis occurs, the two chromosomes in each of your mom's chromosome pairs switch their genes around. That way, your mom doesn't give exactly the same chromosome to two of her children. The same happens in your dad. Each time meiosis happens, then, different patterns of genes occur. The result is that brothers and sisters don't look exactly alike, unless they are identical twins.

Time for Twins?

Since chromosomes get rearranged differently each time meiosis happens, brothers and sisters formed from different gametes will never get exactly the same genes, so they will not look exactly alike. So how do twins happen? Well, there are two kinds of twins: **fraternal** (fruh tur' nul) **twins** and **identical twins**, and they are conceived in different ways.

Fraternal twins develop in the womb together, but they do not look alike when they are born. In fact, sometimes one of the twins is a girl, and one is a boy. This happens when your mom has more than one gamete that is active. Usually, a woman has only one gamete active at any given time, but a man has many. When there is only one gamete from the mother, only one baby can be made. If a mom has two gametes active (it's not something she can control – it just happens), one of the dad's gametes can join with one of the mom's gametes, and another of the dad's gametes can join with the other gamete from the mom. As a result, two separate babies are made, but they are made at the same time, so they develop in the womb together. Since they each came from different gametes, however, they each have different DNA and therefore do not look exactly alike.

These fraternal twins have different DNA.

Identical twins are conceived in a completely different way. Remember that when a gamete from a mother and a gamete from a father combine, the result is a new baby. That baby starts out as a single cell – a stem cell that can become any kind of cell in the body. Well… the first thing that cell does is divide to make two cells, and then those cells divide to make four cells, and so on. For a while, all those cells are stem cells. Sometimes in the early moments of these divisions, some of the cells drift away from each other far enough that they lose contact with the other cells. So now there are two groups of cells in the mother, and they don't know about each other. So they each develop into a baby! Because both groups of cells came from the same original cell, they have the same DNA. As a result, they look the same, and we call them "identical twins." However, while their DNA is the same, they are really not identical. We'll talk about that at the end of the lesson.

These girls are identical twins because they have the same DNA.

Redhead Revelation

So what if your mother has black hair and your father has blond hair, but you have red hair? How could you have received your red hair from either of your parents? Well, some genes are **dominant** and others are **recessive**, and how they interact leads to all sorts of differences between you and your parents. Remember, you have 23 pairs of chromosomes. For each pair, one of those chromosomes came from your mother, and the other came from your father. Your mother's chromosome has her genes, and your father's chromosome has his genes. This means that you have two versions of each gene – one from your mother, and one from your father.

So how do two genes work to give you a trait? Many times a gene is dominant. That means it will determine the trait, regardless of what the other gene is. For example, whether or not you can roll your tongue into a "U" shape when you stick it out of your mouth depends on a single gene in your DNA. If you have the dominant form of that gene, you can roll your tongue. So let's suppose your dad can roll his tongue, and he gave you that form of the gene. Let's suppose your mom cannot roll her tongue, and she gave you that version of the gene. You would have one gene that allows you to roll your tongue, and one gene that does

not. Since the gene that allows you to roll your tongue is dominant, you will be able to roll your tongue. The gene your mom gave you is recessive, or sleeping, so to speak. Of course, it's not permanently sleeping.

Why isn't the gene permanently sleeping? Well, suppose you marry someone who cannot roll his or her tongue. Suppose you have a baby, and your spouse gives the gene that says the baby can't roll his tongue. If you give the baby your mother's gene that says you can't roll your tongue, then your baby would have two genes that say he can't roll his tongue. Because of this, he won't be able to roll his tongue. Since you have both the gene that says you can't roll your tongue and the gene that says you can, whether or not your child can roll his tongue will depend on which of those two genes you give your child. When you have a trait (like red hair) that doesn't belong to either of your parents, it is probably because of a recessive gene that your parents gave to you.

Whether or not you can do this depends on a single gene. If either of your parents gives you the dominant form of the gene, you will be able to do it. If both your parents give you the recessive form of the gene, you will not be able to do it.

Gregor Mendel

The way dominant and recessive genes work was discovered by a monk named Gregor Mendel, who was born in 1822. He was the son of a farmer and became fascinated with plants. While working with pea plants, he noticed that some of the pea plants differed from one another in observable traits. Some pea plants were tall, and some were short. Some had purple flowers, and some had white. Some plants produced green pods, while others produced yellow. It was the same with the peas themselves. Some were green and some yellow. Some of the peas were wrinkled, and others were smooth.

If you took botany, you learned that plants reproduce when the pollen from the flower of one plant is transferred to the flower of another plant. In pea plants, this can actually be done by hand. As a result, Mendel was able to experiment by forcing one plant to reproduce with another plant. One thing he did was to make a short pea plant

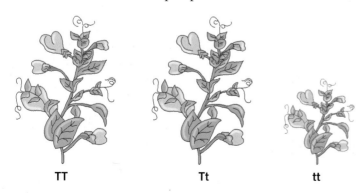

TT   Tt   tt

Mendel said every plant had two factors (genes) for height. If it had at least one factor for tallness, it would be tall. In order to be short, both its factors for height would have to be short.

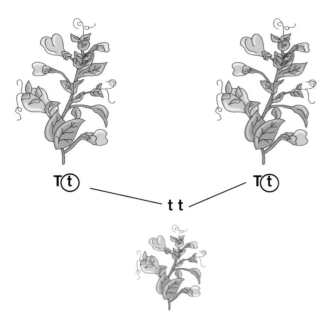

Two tall plants can produce a short plant, if each tall plant has a recessive gene for height and gives it to the offspring.

reproduce with a tall pea plant. Guess what kind of pea plant was made. Did you say a medium-sized pea plant? That's what a lot of people think, but it turns out to be wrong! When a tall pea plant reproduced with a short pea plant, some of the offspring (the new plants) were as tall as the tall parent plant, and some were as short as the short parent plant.

After experimenting with lots of different traits in lots of different pea plants, Mendel came up with an idea – a hypothesis – to explain his results. He assumed that each pea plant had two "factors" that determine its size. Using a capital "T" to stand for the factor that allows pea plants to grow tall and a lowercase "t" to stand for the factor that keeps pea plants short, he said that if a pea plant has two T factors, it will grow tall. If a plant has one T factor and one t factor, however, it will still grow to be tall, since the T factor is dominant. The only way a pea plant would be short is if it had two t factors. Using this hypothesis, he could explain all his results. We use basically the same reasoning today, but we use the word "gene" instead of "factor."

So suppose a tall pea plant reproduces with a short pea plant. Both of the short pea plant's genes must be t, because that's the only way a pea plant will be short. So we could call it a "tt" plant, indicating that both its genes for height will make it short. However, the tall pea plant could either have two T genes (making it TT) or one T gene and one t gene (making it Tt). If the tall pea plant gave a T gene to its offspring, the offspring would be tall, because even though the other parent gave it a t, it would still be Tt, and that would mean a tall plant. However, if the tall plant gave its offspring the recessive t gene, the offspring would be short, because it would be a tt plant.

Just to make sure you understand, think about this. What would happen if two tall plants reproduced? Let's say each plant is a Tt plant. So each plant is tall, because it has the tall gene, but each plant also has the recessive short gene. Well, if each parent gave the offspring the recessive t gene, the offspring would be a tt plant and would be short. So…even though *both its parents were tall, it would be a short plant.* That's why recessive genes aren't permanently asleep. Their traits can show up in the offspring, even though the parents don't have those traits!

This happens with people too! Do you know what a cleft chin is? You have a cleft chin if you have an indentation in the middle of your chin. The gene that tells your body to make a cleft chin is dominant over the gene that says not to make it. So, two parents with a cleft chin can have a child that doesn't have a cleft chin. After all, if each parent has one dominant gene and one recessive gene, they will each have a cleft chin. However, if they each happen to give their child the recessive gene, the child will have two recessive forms of the gene, and the result will be that the child will not have a cleft chin.

This man has a cleft chin, so he has at least one dominant form of the gene.

Do you have any features that neither of your parents has? Ask your parents. Did one of your grandparents have that feature? Then it was probably a recessive gene hidden inside one of your parent's chromosomes. If you were adopted, you may wonder what your biological parents looked like. Chances are you are a mixture of genes from your biological mother and your biological father and their parents. Sometimes, though, you may look like a relative that lived hundreds of years ago! Sometimes parents will have a child that looks like neither of them! If they have pictures of relatives that lived long ago, they might find someone that looks a lot like their child.

There are a number of traits inherited like this. Freckles are dominant, whereas having no freckles is recessive. Dimples are dominant, as is having a white streak in your hair. Free earlobes are dominant; earlobes attached to the side of your head are recessive. The ability to roll the tongue is dominant; the inability is recessive. A longer second toe is dominant, whereas a longer first toe is recessive.

Explain what you learned about genetics.

Personhood

So now you have learned an enormous amount about human anatomy and physiology. All the different organs, cells and other parts we have studied come together to make a human body. But a human body in and of itself doesn't make a person. As you know, you are more than just a body. You are a person with thoughts, ideas, feelings, and memories. You don't live by instinct, like animals. Your actions are usually in response to your thoughts, your beliefs, your memories, and (hopefully) the Holy Spirit guiding you.

Because you have a soul and a spirit, you can pray to and worship God. Animals can't.

In His Image

You see, this is exactly what separates man from all the other creatures God made. Though some science books teach that we are nothing more than advanced animals, this is not so. Every other creature is totally dependent on instinct for survival. Instinct is internal programming to respond in particular ways to certain situations without thinking about it. Here's an example: if an animal is very hungry and comes across food, it will usually try to eat as much of the food as it can to satisfy its need, even if other hungry animals are with it. The animal will not usually consider the hunger of the other animals and try to share the food evenly; its instinct, or internal programming, is to fend for itself. In contrast, let's say you and a friend are hiking in the woods and you both become very hungry. Suddenly you discover there is only one sandwich left. Your conscience tells you to share the sandwich, even if it means your need for food is not completely met. This is because you are a human and not an animal. You wouldn't eat all of the food and leave your companion to starve, because you know that's not right. You make decisions separate from instinct and self-preservation (keeping yourself alive). This is because as a human being, you have a body and soul given to you by God.

God made man in His own image. God formed man from the dust of the ground then breathed into his nostrils the breath of life. (Gen. 2:7). As a result, man became a living body and soul. God said, "*Let Us make man in Our image, according to Our likeness; and let them rule over the fish of the sea and over the birds of the sky and over the cattle and over all the earth, and over every creeping thing that creeps on the earth*" (Genesis 1:26).

Your soul is stamped with the likeness of God. What does it mean that you are created in God's image? This does not mean that God has a human body that looks like a man's, although Jesus took on a man's body in order to come to earth and save us from our sins. The Bible tells us that God is Spirit. So, the image of God that we have within us is not something we can touch, like a human body; it is the likeness of God that's reflected in our inner being: our thoughts, feelings, desires, attitudes, and decisions. Sharing your sandwich with another when you really want the whole thing for yourself reveals you were created in God's image, with the ability to choose right *or* wrong. When you make music, write stories, draw pictures, and come up with great ideas, you are giving evidence of the image of God inside you. Animals cannot do these things. You are special, because you have been created in the image of God.

Because all people are born with this likeness of God, they are able to love, care for, and sacrifice for others. Sadly, because of sin, we live in a world that is now filled with evil. As you know, many choose evil over good. However, we often see people choosing what is right over what is wrong. People will usually help a lost child rather than hurt him. People will usually (but not always) be polite to a stranger. They usually treat their parents and grandparents with respect. If you are a Christian, you have been freed from sin, and you have the ability to choose right over wrong.

Apes and Apemen?

As you may know, some science books teach students the absurd notion that people are just advanced animals simply because people have some things in common with apes. For example: Apes, though they prefer to walk on all fours, can walk upright for a short period of time; however, so can bears. Also, an ape's front feet look similar to human hands, but so do a raccoon's. And finally, human faces have some features in common with ape faces. Because of these similarities, some people think that humans must be related to apes. They think that long ago, creatures that were a lot like apes started having baby apes that had more advanced abilities and these abilities kept advancing over the years. They believe that over many, many, many generations, those advancements "piled up," turning apes into humans.

Although changes do occur in species, such changes have never produced a different creature. For example, many animals long ago were much larger than they are today. But they are still the same animal. The changes in species produce what we call the variety in species. With this variety, we see bigger beaks, longer legs, shorter ears and lots of other feature changes - but the animal does not change into a different kind of creature. The same kind of genes found in the ancestors of the animal are still present in the creature. This is true for all the living things God made: plants, animals and man.

Yet, there are changes that can alter genes. These changes are called mutations. But here's the thing to remember: mutations never result in positive changes; they produce neutral or negative changes. Mutations do not help animals or people. In fact, they often hurt them, causing deformities and diseases, like cancer. However, even though mutations are never positive,

Just because apes and humans have similar features does not mean they are related. God used many similar features when designing life on earth. This points to a common Designer, not a common ancestor.

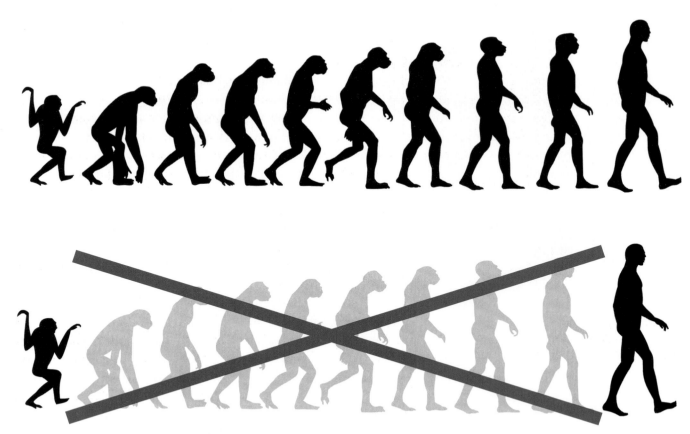

Although the image on top is found in many books, the only creatures that have ever been discovered from fossil remains are the first and the last. All the other creatures pictured in between the monkey and the man have never been found to have existed. However, the drawing itself is what misleads people to believe in evolution, as they do not realize that no evidence for these in-between creatures has ever been found.

evolutionists believe that apes turned into men because of millions of positive mutations piling up over the years! They think these positive mutations changed apes from simple, dumb animals with limited abilities into supremely brilliant human beings with amazing abilities. But in order for the ape to have mutated into a creature that could communicate with words, it would have had to develop a new mouth and tongue. New brain lobes would have needed to develop in order for it to reason and understand complicated concepts. Could an ape really mutate into a creature with the ability to sing, dance, and compose music – the ability to construct cities and engineer computers – the ability to worship God? Frankly, the number of positive, upward mutations needed to do all of this would be so vast, it would be unbelievable. Yet, in nature and in the science lab, even simple positive mutations never occur. Scientific studies have shown over and over again that mutations are never positive for any species. They never cause a species to move upward or become more intelligent. If anything, mutations are negative, making the species less able to survive.

We do not have time to discuss all of the evidence against evolution in this book. However, a great many resources exist that will help you fully understand this issue. Some are listed on the course website. But let's take a minute to learn two very good reasons why a serious scientist should not believe in the theory of evolution.

First, there is no evidence for it. Remember, according to evolutionists, apes were supposed to have evolved into humans long, long ago. Since there weren't scientists around to study it happening, the only way we can see if it happened is to look for evidence of it in the fossil record. Fossils are the preserved remains of things that were once alive. So, if there were once creatures that were becoming humans, there should be fossils preserved from all those different creatures. Do you realize that absolutely no fossils exist of a half-man/half-ape. There are simply no apeman fossils.

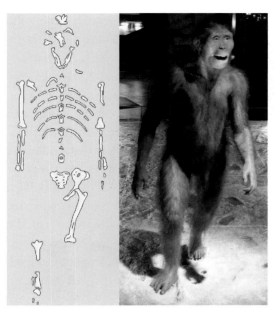

From the fossil remains on the left, the museum created this imaginary creature and called it Lucy. No such creature has ever been found to have existed.

Take a look at the images here. The first image is a drawing of the actual fossilized remains of a creature that some scientists think was a part of the evolution between ape and man. The second is a model of what the scientists think this creature looked like. Notice that the model has the creature walking upright, with a face that looks a bit human, and hands that look quite human.

Take another look at the actual fossils that were found. Do you see how few fossils there are? The fossils found could not give us that much information about the creature. There is only a tiny portion of the skull; so how could they know what the face looked like? Only a tiny fraction of hand bones are in the fossil. How can they make human-like hands when they only have a few small bones? Also, notice that no hair was fossilized, but the model has this creature partly covered in hair, misleading people to believe that this was a hairy man-like creature.

Scientists made the creature look as they wanted it to look, rather than waiting until they had enough fossils to determine how it might have really looked. Look at the fossil from this creature again. You already noticed how few bones were found in this fossil. Well, guess what! This is one of the most complete fossils of all the creatures that are supposed to be a part of the evolution of apes into people. Most of the drawings of the imaginary "apemen" that you see in books or museums were based entirely upon a tiny handful of bones! No fossil evidence exists for the idea that people evolved from apes! That is one important reason not to believe in evolution.

Secondly, just because things may have similarities, doesn't mean they are related. Look at the two pictures here. One is a bumblebee on a flower. The other is a flower called a bee orchid. They look very similar but are not even in the same biological Kingdom. One is a plant; the other is an animal. But here's something to think about: When you draw pictures, do you tend to create similar drawings? I can always tell which of my children drew something because I can identify the unique characteristics found in their drawings. If you study famous artists, you might be able to tell who painted a particular

The bumblebee (right) and the bee orchid (left) look similar, but are not related. Simply because creatures look similar does not mean we can assume they are related.

painting just by looking at it because artists use similar designs and patterns in their works. In the same way, the world was created by a Designer who used similar patterns and designs in His work! The similarity between apes and people occurs because they were both made by the same Designer – the Lord God.

What about Cavemen?

What about all the evidence that humans lived in caves at one time? Doesn't that mean at one time people were really stupid (like apes) and didn't know how to build things? Well, even today, there are people who live in caves. It's evident that when people from long ago would wander from place to place or move to a new location, they would spend the winter in a protected environment. Caves are perfect places for this. Caves can be huge with many rooms and often have natural springs with fresh water. Perhaps some families decided they preferred

This is an example of the wonderful artwork you can find in caves. It clearly wasn't done by an "apeman."

cave living. That wouldn't be too unusual. Even the Bible has many references to people living temporarily in caves. For example, Genesis 19:30 tells us that Lot lived in a cave for a while, and 1 Samuel 22 tells us that David lived in a cave for a time.

Long ago, when scientists found tools in caves and drawings on cave walls, they immediately assumed those inhabiting the caves were "apemen" that did not know how to build their own homes. However, just because someone lived in a cave and decorated the walls with artistic drawings does not mean the person was not fully human. In fact, the artistic skill of some of the drawings on cave walls shows the shading, use of light and shadows, and use of color that one might find with talented artists today. For an artist to display his art on a stone wall is even more remarkable! Cavemen were cave-dwelling men, not apemen.

Why Did God Create Me?

So you aren't some highly-evolved ape. You are a special creation of the Almighty God. You might ask, "Why did God make me?" "What is my purpose for being alive?" Those are important questions that people have been asking since the beginning. The answer is rather simple: the Bible tells us that the reason God made you is so you could bring glory to God through your life and experience joy in a relationship with Him both here on this earth and eternally in heaven.

How do you bring glory to God? Well, everything you do can be done for God's glory if your heart is focused on Him. When we choose to glorify God with our lives, every task we do can bring God glory, whether fixing a doorknob, washing the dishes, singing at church, dancing ballet, participating in a sport, or writing a letter to a friend. I remember watching a movie about the great missionary to China, Eric Liddle. He was preparing to run in the Olympics. His father explained to him that even running can be done for God's glory. He said, "Son, you can give God glory by peeling a potato if you peel it to perfection…Run for God, and let the world stand back and wonder." It is the attitude of our hearts that determines whether we are truly living our lives for God's glory. As Colossians 3:17 tells us, "*Whatever you do, whether in word or deed, do it all in the name of the Lord Jesus, giving thanks to God the Father through Him.*"

God has planned something special just for you! As Ephesians 2:10 says, "*For we are His workmanship, created in Christ Jesus for good works, which God prepared beforehand so that we would walk in them.*" What do you love to do? What interests you? Is there something you particularly enjoy? Answering these questions can help you discover God's plan for your life. Has anyone ever commented that you were

God created you with special talents. You should pray to find out what His plans are for your life.

gifted or talented at something? Well, those gifts and talents were given to you by God. Even your interests and passions are from the Lord. These are what make you different and special. All of these things can be used in your life to glorify God.

If you have not yet discovered your special gifts, talents or interests, ask the Lord to reveal them to you and show you how to use them for His glory. Maybe you have a special interest in animals or inventing things. Perhaps you are talented at learning languages. My brother didn't realize until he was an adult that God had given him a special gift for learning languages. Though he only spoke one language growing up, by the time he graduated from college, he could speak five languages fluently! Imagine if he had discovered this gift earlier! If you ask God, He will guide you along the special path He has chosen for you. Psalm 48:14 reminds us, "*For such is God, our God for ever and ever; He will guide us until death.*"

Studying God's Word and prayer help you grow spiritually.

Grow in Wisdom

As you begin to develop your gifts and passions and find the special path the Lord has for your life, it's very important that you not only grow outwardly (in your physical body), but that you grow inwardly (in your spiritual life). But what does it mean to grow spiritually? It means to grow in godliness, to become more like Jesus. This growth happens as you develop a closer walk with God. Because you live in this world, you cannot keep your body healthy and strong forever. It will one day get old and wear out. But one part of you will never die – your spirit. So, it's important that you keep your spirit strong and healthy. The Bible tells us, "*Though our outer man is decaying, yet our inner man is being renewed day by day*" (2 Corinthians 4:16).

The most important thing you can do on this earth is prepare for your eternal life in heaven. It will last forever. I remember reading in a devotional book that our lives are like a dot with a long line going out from it that spans the entire page. The dot represents your life here on earth. The line represents your life in heaven. Most of us are focused on the little dot, but the really important thing is to focus on the long line – which represents your real life that will go on and on and on. Are you living for the dot or for the line? Are you focused on your short life in this body, or on the long life you'll live in heaven with God?

We've talked a great deal throughout this book about getting exercise to keep your body healthy. Yet, there is another kind of exercise that is even more important. The Bible tells us, "*For physical training is of some value, but godliness has value for all things, holding promise for both the present life and the life to come*" (1 Timothy 4:8 – NIV). What happens to your physical body when you don't feed yourself healthy food or get exercise? Just like your body can grow weak and sick when you make bad choices, your spirit also needs the right things to grow strong and healthy. What does it mean to be spiritually strong? A person that has a strong spiritual life has a personal relationship with God, in which he communicates with God daily – through prayer, His Word, and listening to the leading of His Holy Spirit. This person will do the right thing, even when it's hard. This person will also immediately know when he has done wrong – because he is close to God and can sense God's Holy Spirit. A spiritually strong person will also have more joy and peace than others, because those are things God loves to give to His children. As a Christian, you have a choice to develop a strong spiritual life or not.

How do you grow spiritually? There are many ways to grow, but the most important is to spend time alone with the Lord in prayer and in His Word. At the end of this lesson, you will create a prayer journal where you can begin to write your prayers down while you pray. Another thing you should do is begin to read and study God's word. We'll have some suggestions for doing this at the end of this lesson as well. Also, it's important to spend time with other Christians. We call this fellowship. It helps you grow spiritually, because other Christians can help keep you on the right track. If you have a friend, brother, or sister with whom you can pray and talk to about God, you are quite blessed indeed. If you don't, pray for one. I believe God will bring you a spiritual friend if you ask Him.

I hope you enjoyed your study of anatomy and physiology. You know now how to keep your physical self healthy and strong and also how to keep your spiritual self healthy and strong. You have truly grown in many ways this year – but don't stop! Like Jesus did when he was your age, keep growing!

"And Jesus kept increasing in wisdom and stature,
and in favor with God and men" (Luke 2:52).

What Do You Remember?

How do cells divide? At what point were you alive when you were in your mother's womb? How many chromosomes do people have? What is a trait? What are genes? Where is your DNA located? What is mitosis? What is meiosis? What are gametes? Who was Gregor Mendel? What makes humans different from animals? What are the two main reasons we said you shouldn't believe humans evolved from apes? Who were the cavemen? Why did God create you?

Notebooking Activities

Write down all the fascinating facts that you found interesting about growth, development, and genetics. Draw the stages of development that are shown on pages 233 and 234. Now ask your parents if they have any ultrasound pictures of you when you were in your mother's womb. These pictures aren't ideal, but they give you some idea of what you actually looked like while you were developing.

Possible Purpose Page

At the top of a separate sheet of paper, write out Ephesians 2:10. Then draw a vertical line dividing the page into two parts. On one side of the paper, write down all the special talents or gifts you think you may have. If you're unsure, ask God to show you and write them down as He does. Your parents can also give insight into these things. On the other side of the paper, write down your special interests. These can include: things that are exciting to you, things you know a lot about, things you would love to learn more about, and things you remember easily. At the bottom of the page, write down how you can use these talents and interests to glorify God with your life!

Prayer Journal Activity

I hope you are inspired to work on your prayer life! If you are, developing the habit of keeping a prayer journal is a great idea. If you begin this habit now while you are young, it will help keep your walk with God strong throughout your high school and college years. Even if you have trouble writing by hand, that's okay. No one will see your prayer journal, and God is most concerned about your heart. Just keep working your hand muscles as you jot down your prayers each day. Before you know it, the art of handwriting will become easier and easier!

You will need a blank notebook, journal, composition book, or spiral notebook. Whatever you choose to use will be just fine. When you spend time with God each day, write your prayers down as you pray. This will help keep your mind focused on praying so you don't begin thinking about lunch or your next birthday. It's easy to get distracted, but since writing takes so much longer than thinking, you'll pray longer and spend more time with God! Also, you can go back and read your prayers months from now and be amazed at the ways God answered them. Sometimes He answers immediately; sometimes He waits awhile; other times He has a better plan. Keep praying! You'll see!

Bible Reading Plan

Although you may have heard a lot of Bible stories in your lifetime, you'll be surprised at what you learn when you actually read the Bible for yourself! Personally reading the Bible each day will help grow your spiritual life tremendously. The Bible is God's Word, God's love letter to us. He wants us to know Him, and that's the reason He gave us the Bible. Through reading the Bible, we get to know God better and understand Him more. The Bible is not just any old book; it's called the "living Word of God," because the Holy Spirit enables us to understand deep truths when we read it. As we read, the Holy Spirit will lead us and show us many things we need to understand in order to grow strong spiritually. Hebrews 4:12 tells us, "*For the word of God is living and active and sharper than any two-edged sword, and piercing as far as the division of soul and spirit, of both joints and marrow, and able to judge the thoughts and intentions of the heart.*"

Before you read your Bible each day, be sure to ask the Holy Spirit to guide your heart as you read, so you can understand what God is saying to you. Reading the Bible is sometimes hard without a plan. If you go to the course website we told you about in the introduction to this book, you will find some websites that help you plan how much of your Bible you will read every day.

Project
Dominant and Recessive Traits

You learned that genes are dominant and recessive, and certain traits that people have are determined by just one gene. The gene that produces freckles is dominant, whereas the gene that produces no freckles is recessive. The gene that causes you to have a dimple when you smile is dominant, and the gene that keeps you from having a dimple is recessive. The gene that keeps your earlobes free from the side of your face is dominant, while the gene that makes them attached is recessive. The gene that gives you the ability to roll your tongue is dominant, while the gene that keeps you from being able to do that is recessive. The gene that makes your second toe longer than your first is dominant, whereas the gene that makes the first toe longer is recessive.

In this project, you need to examine your family members for these traits. Look at your parents, brothers, and sisters and examine each trait listed here: **freckles, dimples, attached/free earlobes, ability to roll the tongue, and length of the first two toes**. Try to determine what possible genes each of your family members could have based on what you observe. Then, see if you can figure out how the genes were passed from your parents to you and your siblings.

Here's an example of what you need to do. Look at each member of your family's earlobes. For most people, their earlobes are either attached or free, as shown on the next page. As you have been told, the gene for a free earlobe is dominant, so the gene for an attached earlobe is recessive. The only way a person can have an attached earlobe is if that person has two of the recessive genes. However, if a person has just one of the dominant genes, that person's earlobe will be free. If we call the gene that makes you have a free earlobe "F" and the gene that makes you have an attached earlobe "f," that means a person with an attached earlobe must have two "f" genes (ff). However, a person with a free earlobe might have just one "F" gene (Ff), or the person might have two (FF).

Now you need to look at your mom's ear. Is the earlobe attached? If so, your mom has two "f" genes (ff). Write that down. Is it free? If so, you don't know exactly what genes your mom has. It might be just one "F" (Ff) or two "F" genes (FF). Write down both possibilities. Do the same for your dad. Now look at your own ears in the mirror. Are the earlobes attached or free? Use the same kind of reasoning and write down what genes you might have. Now look at your parents' genes and your genes. Can you come up with the possibilities of which genes your mom and dad gave you?

Let's go through an example to help you understand what I mean. Let's say your dad has a free earlobe. That means he could be either FF or Ff. Let's say your mom has an attached earlobe. That means she is ff. Now let's say that you have a free earlobe. What does that mean? Well, you could be either FF or Ff, because you need only one dominant gene to have a free earlobe. However, you can't be FF, can you? Remember you get one of these genes from your dad and one from your mom. Your mom has only two "f" genes. As a result, she can only give you an "f," which means one of your genes has to be an "f." That means you must be "Ff." Now think about that for a minute. Without any fancy equipment or laboratory, you can actually determine what genes are on your DNA!

Of course, this only works if things work out right. If both your parents have free earlobes, they could both be either FF or Ff. If you also have a free earlobe, you could be FF or Ff. Since each of your parents has each of the genes, you can't really determine which genes you have. Since either of your parents could have an "f," either of them could have given you an "f," and since it is recessive, you will have no way of seeing it.

The goal of this project, then, is to see if you can determine the genes you have for any of the traits listed in bold at the beginning of the experiment. Also, try to figure out if you can determine the genes your brothers and sisters have for the same trait.

There is one other thing you should look for when you are looking at all these traits. Do you remember that early on in this lesson (p. 235) I told you that your genes are not the only consideration when it comes to your traits? Your environment plays a role as well. Remember, for example, that identical twins have identical DNA, but they are not identical. They have different fingerprints, for example. Why? Because they have slightly different environments. Believe it or not, just because they were in slightly different places as they developed in their mother's womb, their fingerprints developed differently.

As you look at the different traits, notice that the details of the traits vary from person to person. For example, if your mom has freckles, that means she has at least one dominant gene for freckles. If you have freckles, that means you have at least one dominant gene for freckles as well. However, the number of freckles you have might be quite a bit more than the number of freckles your mom has. That's because while genetics controls *whether or not* you have freckles, the *number and prominence* of the freckles is affected by your environment. This is a very important point. While most of your traits are rooted in your genetics, most of them are also strongly affected by your environment. The kind of person you end up being, then, really depends on both. Your genes play a role, but also how and where you live, what you eat, and who you end up spending time with will also play a role. Both are important, and hopefully you will see that as you examine traits in this project.

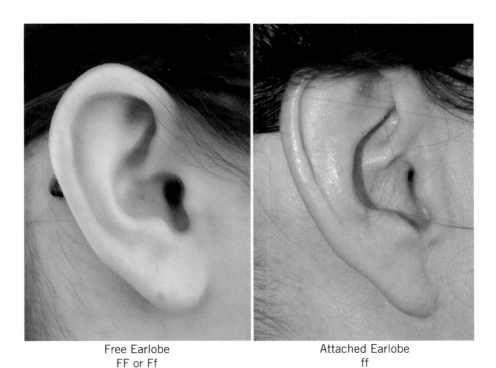

Free Earlobe
FF or Ff

Attached Earlobe
ff

Answers to the Narrative Questions

Your child should not be expected to know the answer to every question. These questions are designed to jog the child's memory and help him put the concepts into his own words. The questions are in plain type. *The answers are highlighted in bold and italic type.*

Lesson 1

What tells us that the Egyptians understood a lot about anatomy? *The way they mummified people.* How do the laws that God gave to the Hebrews show us that God cares about our health? *God's laws protected people from germs they could not see.* What was wrong with the way the Greeks decided on their scientific beliefs? *They did not test them.* What did Galen use to treat the gladiators' wounds? *He used clean rags and wine-soaked rags.* What did Hooke call the tiny rectangles he saw in the cork he examined under a microscope? *He called them cells.* Name the different cell parts about which you've learned. *Cell membrane, mitochondria, lysosomes, Golgi bodies, ER, centrioles, and the nucleus.*

Lesson 2

Name the different things that bones do for your body. *They give you form and hold you up, they protect organs, they make blood cells, they store minerals and fats, and they help you move.* What is the periosteum? *The outer layer on a bone that has nerves and blood vessels.* What mineral makes compact bone strong and hard? *Calcium.* What is the bone tissue that forms tunnels and pores called? *Spongy bone.* What are the two kinds of bone marrow? *Red bone marrow and yellow bone marrow.* What are osteoblasts? *The cells that make new bone.* Where are the smallest bones in your body found? *In the ear.* Which is the longest bone in your body? *The femur.* What do ligaments do? *They hold the bones together.* What is the rounded part of your skull called? *The cranium.* Can you name at least five other bones in your body by their scientific names? *Look at the diagram on page 45 to confirm the answer.* Can you name at least one kind of joint? *Look at pages 51 and 52 to confirm the answer.*

Lesson 3

What are the three kinds of muscle tissue in your body? *Skeletal, Smooth and Cardiac.* Which of those muscle kinds are voluntary, and which are involuntary? *Skeletal muscles are voluntary, smooth and cardiac are involuntary.* What is muscle tone? *Partial contraction of your muscles throughout the day.* What are tendons? *The tough fibers that connect your skeletal muscles to your bones.* Where is your Achilles tendon? *In the back of your ankle.* What are antagonistic muscles? *Muscles that work opposite of each other.* What do muscle cells have a lot of that give them energy? *Mitochondria.* What things did we discuss that help muscles move? *Glucose, oxygen, vitamins and minerals.* What do muscles need to grow? *Protein and exercise.* How do you keep your cardiac muscle strong? *By elevating your heart rate for 30 minutes or more several times a week.* Name two places in your body where you find smooth muscles. *Answers will vary but might include blood vessels, stomach, intestines, and bladder.*

Lesson 4

What is the white outer layer of your tooth called? *Enamel.* What is the layer right below that called? *Dentin.* What is the hardest substance in your body? *Enamel.* Name a few things saliva does for you. *Begins chemical digestion, protects teeth, defends against infection, and makes food taste better.* What is the name of the pipe that food goes down after you swallow it? *Esophagus.* How do your stomach and esophagus keep from getting burned by your own stomach acid? *Mucus is produced in the stomach, and the gastroesophageal sphincter keeps it in the stomach.* What is the food called when it enters the small intestine? *Chyme.* What happens in the small intestine? *Digestion is completed, and nutrients are sent to the bloodstream.* Which organ is like a huge chemical factory? *Liver.* What do the kidneys do? *They remove excess salts, water, and other substances from the blood and put them in the urine.*

Lesson 5

How do you know if you are dehydrated? *Thirst, dark urine.* Why do simple carbohydrates give you quick bursts of energy, while complex carbohydrates don't necessarily do that? *Many complex carbohydrates take longer to digest.* What do carbohydrates change into inside the body? *Simple carbohydrates, mostly glucose.* Proteins are made of what kind of molecules strung together? *Amino acids.* What is a complete protein? *A complete protein contains all the essential amino acids.* Which foods provide your body with Omega 3 fatty acids? *Salmon, tuna, sardines, dark leafy green vegetables, avocados, sesame seeds, walnuts, pumpkin seeds, flax seeds and sesame seeds.* Name three vitamins that are important to get, and tell why they are important. *Answers will vary.* Where are minerals found? *In the earth and in the world's lakes, rivers, streams, and oceans.* Name two minerals, and tell why they are important for your body. *Answers will vary.*

Lesson 6

What does the hair in your nose do? *It filters large particles out of the air you breathe.* What does the mucus in your nasal passage do? *It captures particles out of the air you breathe.* What are cilia? *They are like motorized whips that push mucus out of your nasal cavity, trachea, bronchi, and lungs.* Explain how the conchae help warm and moisten the air you breathe. *They disrupt the air flow, making the air bounce around and hit the warm, moist tissue in your nasal cavity.* What are the thin strips of tissue in your larynx called? *They are your vocal cords.* What determines your voice's pitch? *The tension in the vocal cords as well as how thick and heavy they are.* What determines your voice's volume? *How much air passes across your vocal cords determines the volume.* How do the cartilage rings around your trachea help you? *They keep your trachea from collapsing.* What are your bronchi? *They are the tubes that carry air from your trachea to your lungs.* Explain the importance of alveoli. *They are the places in the lungs where oxygen is put into the blood and carbon dioxide is removed.* Name some of the dangers of smoking. *It increases your chance of lung cancer, raises your blood pressure, damages your lungs, and is addictive.* How does the oxygen get from your lungs into your blood? *At the alveoli, the oxygen travels across the single-cell-thick blood vessel walls into blood cells.* What muscle is mostly responsible for your breathing? *The diaphragm is mostly responsible for breathing.*

Lesson 7

Why did Jesus give Himself as the sacrifice for our sins? *Because blood must be shed for the forgiveness of sin. Jesus was the perfect sacrifice, so His blood covers all our sin.* What does a person have to do to be forgiven for their sins once and for all and receive life everlasting? *The person must believe that Jesus died on the cross for his or her sin and accept the forgiveness that comes from Jesus' blood.* What are the four functions of blood? *Transportation, Temperature Regulation, Protection, and Message Carrier.* Name the four basic components of blood. *Plasma, red blood cells, white blood cells, and platelets.* What should you do if you are bleeding seriously? *Elevate the wound, apply pressure to the wound, and get medical care.* Where are your blood cells made? *In the red bone marrow.* Why is it important to give people the right type of blood? *If the person's body recognizes the new blood as foreign, it will attack the blood cells.* What are the four blood types (not including the Rh factor)? *A, B, AB, and O.*

Lesson 8

What do we call the top two chambers of your heart? *Atria.* What do we call the bottom two chambers? *Ventricles.* What are the veins leading from the lungs into the heart called? *Pulmonary veins.* What are the arteries leading from the heart to the lungs called? *Pulmonary arteries.* What is the main artery that takes blood out of the heart to the body? *Ascending Aorta.* What are the names of the two veins that bring deoxygenated blood from the tissues of the body back to the heart? *Superior Vena Cava and Inferior Vena Cava.* What is it that you are hearing when you hear your heart beat? *You are hearing the valves closing.* What do the two numbers in a person's blood pressure mean? *The first number is the pressure against the artery walls when the ventricles contract (systolic blood pressure). The second number is the pressure against the artery walls when the ventricles relax (diastolic blood pressure).*

Lesson 9

Name the three parts of the brain we discussed. *Cerebrum, cerebellum, brainstem.* What part of a neuron receives information? *The dendrite.* What part sends the information on its way? *The axon.* Where do you find the nucleus in a neuron? *In the cell body.* What does a sensory neuron do? *It takes information to the brain or spinal cord.* What does a motor neuron do? *It sends information from the brain or spinal cord to the muscles and organs of the body, telling them what to do.* What is the somatic nervous system? *It is the part of the nervous system responsible for the voluntary movements you make.* What is the autonomic nervous system? *It is the part of the nervous system that covers the involuntary activities of the body.* What does the endocrine system do? *It makes and releases hormones to control activities in the body.*

Lesson 10

Name the four lobes in the cerebrum. *Frontal, parietal, temporal, and occipital.* Which lobe is responsible for speech and language? *Frontal.* Which lobe is responsible for your emotions? *Frontal.* Which lobe processes smell and memory as well as tone and loudness? *Temporal.* Which lobe processes sensory information and integrates it to determine where you are in relation to your surroundings? *Parietal.* Which lobe processes visual information? *Occipital.* Which side of your brain is more active when you are doing math? *The left side.* Which is more active when you are working on a piece of art? *The right side.* How does myelin help your neurons? *It increases the speed with which signals can travel.* Which part of the brain is responsible for keeping you balanced? *Cerebellum.* What is the reflex arc? *It is a sensory neuron, an interneuron in the spinal cord, and a motor neuron. If the sensory neuron senses danger, the interneuron sends a signal to the motor neuron, which causes you to quickly move away from the danger without your brain being involved.* Why are interneurons and interconnections between neurons important? *It is how we learn and remember things.*

Lesson 11

What is the part of your nose that holds your olfactory cells called? *Olfactory epithelium.* How are odors received and transferred to the brain? *The odor is received by cilia of the olfactory cells and then transferred to the olfactory bulb, which sends the information to the brain.* Where are your taste buds found? *In the papillae of the tongue.* What are the five taste sensations? *Sweet, sour, salty, bitter, and umami.* What is the pinna? *It is the part of the ear you see from the outside.* Why do ears make wax? *The wax traps foreign particles to keep them from getting to the middle or inner ear.* Name the three bones in the middle ear. *The malleus (hammer), incus (anvil), and stapes (stirrup).* What are otoliths? *Tiny stones in the ear that help you with balance.* What is the sclera? *It's a tough, thin bag that forms your eyeball.* What is the pupil? *It is the part of your eye through which light enters.* What is the iris of an eye? *The part of the eye that adjusts how much light gets in.* What is the fovea? *A spot on the retina where the cones are concentrated – the center of your vision is focused there.* What cells enable you to see in color? *Cones.* What cells enable you to see in dim light? *Rods.* Name some of the ways God added special protection for your eyes. The bones that protect your eyes are very strong, your eyelids blink to protect your eyes, the conjunctiva protects the white part of your eye, and your tears protect your eyes.

Lesson 12

What is the system called that includes your skin, nails, and hair? *The integumentary system.* How do skin, nails, and hair grow? *They grow at the base, and the new cells push the old ones away from the base.* What are the two layers of the skin? *The epidermis and the dermis.* What is the layer of tissue underneath those two layers? *The hypodermis.* What two important pigments are mostly responsible for skin color? *Melanin and carotene.* What do sebaceous glands make? *They make sebum.* What does sweat do? *It cools your skin.* What are the three layers of a hair? *The cuticle, cortex, and medulla.* Which layer is sometimes missing? *The medulla.* Which layer of skin is responsible for making your fingerprints? *The dermis.* In both hair and nails, what is the name of the region in which new cells are made? *The matrix.*

Lesson 13

What is a pathogen? *It is an organism that can invade your body and do harm.* What is an infectious disease? *It is a disease caused by an infection of some kind.* Name one kind of disease. *There are a lot of answers here – the flu, a cold, cancer, heart disease, polio, etc.* How do lymph nodes protect us from diseases? *They filter the lymph to remove pathogens.* Where can stem cells that develop into disease-fighting cells be found? *The bone marrow.* Name some of the systems that help keep us well as our first line of defense. *Skin, sebum, tears, mucus, cilia, helpful bacteria.* What name is given to that part of our immunity? *Innate immunity.* What does a fever do? *It increases chemical activity in your body and makes it more difficult for many invaders to survive inside your body.* What do we call the immunity we get from cells that analyze attackers and remember them in case they come back? *Adaptive immunity.* How are antibodies formed in the body? *They are made by B cells.* How do vaccines work? *They give the body a weakened or inactive pathogen (or just an antigen from a pathogen), stimulating adaptive immunity.* How was penicillin discovered? *It was discovered when Fleming noticed that a mold he was growing killed bacteria.*

Lesson 14

How do cells divide? *They can divide by mitosis, where one cell turns into two cells that are the same as the original, or they can divide by meiosis, where one cell turns into four cells that each has half the DNA of the original.* At what point were you alive when you were in your mother's womb? *As soon as your mother's gamete joined with your father's gamete.* How many chromosomes do people have? *They have 46 chromosomes, which come in 23 pairs.* What is a trait? *It is an observable characteristic, like eye color.* What are genes? *They are the parts of the DNA that help determine your traits.* Where is your DNA located? *In the nucleus of almost every cell.* What is mitosis? *It is where one cell copies its DNA and makes a total of two cells, each of which is identical to the original cell.* What is meiosis? *It is where one cell copies its DNA and makes four cells, each of which has half the DNA of the original.* What are gametes? *Reproductive cells that have only half the DNA of normal cells.* Who was Gregor Mendel? *He was the monk who discovered the basic laws of genetics.* What makes humans different from animals? *We are made in God's image.* What are the two main reasons I said you shouldn't believe humans evolved from apes? *There is no fossil evidence for it, and similarity doesn't always mean relatedness.* Who were the cavemen? *They were people who lived in caves. They were not apemen.* Why did God create you? *God made you so you could bring glory to God through your life and experience joy in a relationship with Him both here on this earth and eternally in heaven.*

Photograph and Illustration Credits

Photos by Jeannie Fulbright: 19, 20 (top), 34-35 (all), 37 (both), 40 (bottom), 48 (middle), 52 (bottom), 53 (all), 55, 56 (top), 61, 63, 66 (bottom), 67-68 (all), 82-83, 94, 100, 106 (left), 109, 117, 118, 124, 132, 136, 140, 143 (both), 146-147 (all), 151 (top), 154 (top), 156 (bottom left and right), 159, 164, 165 (bottm), 180, 195, 198, 207 (right), 209

Photos and illustrations from www.shutterstock.com: (copyright holder in parentheses): 21 (Carolina K. Smith, M.D - bottom), 22 (Georgios Alexandris), 32 (Matthew Cole – bottom), 33 (Smit), 51 (Nikolay Okhitin), 54 (Aquatic Creature), 57 (Alexonline – top), 58 (Simone van den Berg), 62 (Tracy Whiteside -bottom left and right), 64 (PeterG), 65 (Digitallife), 70 (Tara Urbach), 72 (Simon Krzic), 74 (Jubal Harshaw - bottom), 80 (Axel Kock), 81 (Wolfgang Amri - top), 92 (Emanuele Tortora - bottom), 97 (Jens Mayer), 101 (Julián Rovagnati), 103 (Anna Jurkovska), 112 (Radomir Jirsak - top), 113 (Ciapix), 115 (Patrimino Designs Limited - top), 125 (AnantRohankar), 126 (Ovidiu Iordachi - top), 130 (Timothey Kosachev - bottom), 138 (Hkann - bottom), 142 (Niderlander), 146 (GeoM - top), 145 (Alexander Vasilyev), 149 (Matthew Cole), 151 (Chris Harvey – bottom), 153 (Andrea Danti), 157 (Mandy Godbehear - bottom), 162-163 (NL Shop - all), 165 (Kristiana007 - top), 171 (Erzetic - top), 171 (Darryl Vest - bottom), 175 (Megastocker – balance icon from www.clipart.com), 176 (Ismael Montero Verdu - bottom), 178 (Elena Ray), 180 (Christopher Meade – top), 184 (Blamb), 186 (Mmutlu - top), 189 (Dyonisos Design), 190 (Martin Valigursky - top), 190 (Steve Snowden - bottom), 192 (Dmitriy Shironosov), 196 (Andrea Danti), 197 (Jubal Harshaw), 199 (Margot Petrowski - right), 201 (Andrea Danti – right top and right bottom), 201 (Jubal harshaw – bottom left), 202 (Cheryl Casey), 212 (Andrea Danti – right), 212 (Jubal harshaw – left), 206 (Morozova Tatyana [Manamana]), 207 (Oguz Aral - left), 208 (Blamb), 210 (Dan Ionut Popescu – left fingerprint), 210 (Charobnica – middle and right fingerprints), 211 (Tootles), 216 (Fedor Kondratenko), 222 (Rob Byron), 224 (Blamb - top), 224 (Catalin Petolea), 235 (Ronald Sumners), 227 (Christopher Meade), 229 (Sven Hoppe – left), 231 (Lukas Pobuda – text added by Kathleen Wile), 232 (Alena Ozerova), 235 (Emin Kuliyev), 237 (Ykh), 240 (Cheryl Casey - top), 241 Les3photo8 (bottom), 242 (Zdorov Kirill Vladimirovich), 243 (Eric Isselée), 256 (Risteski Goce – top two images), 246 (Lolloj), 247 (Tara Flake)

Photos from www.clipart.com: 20, 21 (top), 28 (top), 30 (bottom), 31 (top), 81(bottom), 88 (top left, bottom right and left), 91, 98 (bottom), 99, 120, 175 (balance icon only), 215, 239 (both)

Photos from www.istockphoto.com: (copyright holder in parentheses): human form image on page and chapter headers (Max Delson Martins Santos), 3 (Chris Lemmens), 85 (Karen Town), 88 (Andrey Yanevich - top right), 94 (Denis Pepin - top), 105 (Doug Berry), 128 (Carmen Martínez Banús), 173 (iofoto - top), 177 (Carrie Bottomley), 185 (Jason Lugo), 240 (Sage78 – right), 245 (Anne-Britt Svinnset - top right), 245 (Andrew Howe - top left)

Photos from www.dreamstime.com: (copyright holder in parentheses): 38 (Spooky2006 – top), 39 (Stephen Mcsweeny – bottom), 86 (Monkeybusinessimages – bottom), 87 (Gregait), 96 (Alkir)

Photos in the public domain: 23, 24 (both), 25, 40 (courtesy of NASA - top), 69 (Mariana Ruiz Villarreal), 95 (courtesy of the CDC), 119, 123 (all three), 127, 133, 193, 201 (left), 217-218 (courtesy of the CDC), 221 (courtesy of the CDC - both), 226 (courtesy of the CDC), 244 (bottom right), 245 (bottom left)

Photos published under the Creative Commons Sharealike License (http://creativecommons.org/licenses/by-sa/3.0/): 128 (Attribution: Clockface), 176 (Attribution: Patrick J. Lynch - left), 186 (Attribution: Patrick J. Lynch - right), 199 (Attribution: Bertas la Blanquita - left)

Index

cells 32–33
cochlear, 183
cranial, 152
dental, 71
motor, 154, 197
muscle, 63
olfactory, 177, 181
optic, 186–187, 189
sensory, 154, 169, 197, 212
skin, 197, 201
spinal, 152, 167–168
nerve endings, 201, 206–207
nervous system, 32, 47, 149–157, 161–171
autonomic, 156–157
central, 149, 152, 156, 233
peripheral, 149, 152
somatic, 155–156
neurons, 151–156, 161, 165, 170–171
afferent, 176
motor, 154–156
sensory, 154–156, 177, 180, 196, 207
neurotransmitters, 153
night vision, 187
nuclear membrane, 29
nucleolus, 30
nucleotides, 30
nucleus, 26, 29–31, 152, 154
nutrients, 27, 40, 70, 76–77, 86, 123, 137–138

O

occipital 163
bone, 46
lobe, 162, 165
olfactory system, 176–177
organelles, 27–29
oropharynx, 107–108
ossicles, 183
osteoblasts, 42–44, 234
osteoclasts, 43
osteoporosis, 39–40
oxidation, 96
oxygen, 61, 63–64, 86, 104, 111–113, 121–122, 125, 134–141

P

pacemaker, 142
Pacinian receptors, 207
palate, 73
pancreas, 78
papillae, 179–180
parietal lobe, 162–163
patella, 45
pectoral girdle, 48
pelvic girdle, 49
pepsin, 75
pericardium, 140
periosteum, 41, 44
phagocytes, 126
phalanges, 45, 49–50, 52
pharynx, 69, 107–108
pinna, 181
pituitary gland, 158
plasma, 124–125
platelets, 123, 127–129
premolars, 70
protein, 31, 60, 90, 125, 198, 222
dietary, 28, 65, 70, 76, 78, 86, 90–91
puberty, 235
pulmonary artery, 139–141
pulp, 71
pulse, 143
pupil, 186–187
pyloric sphincter, 76, 157

R

radius, 45, 49
recessive gene, 239–242
rectum, 78–80
reflex arc, 169
reflexes, 58, 169
REM sleep, 57
remodeling, 41, 44
renal system, 80–81
reproduction, 238,
cell, 29, 218
respiratory system, 32, 103–118
retina, 186–189
Rh factor, 130–131
rib cage, 38, 48, 114
ribosomes, 29–30
rickets, 40, 96, 98

RNA, 29–31
rods, 186–187

S

sacrum 45, 47, 49–50, 167
saliva 71, 72, 180
scab, 44, 128
scapula, 45, 48–49
scar, 205
sclera, 186
scurvy, 95
sebum, 200–201, 221
semicircular canals, 182, 184
semilunar valves, 142
senses, 154–155, 163, 175–191
sensory neurons, 154–155, 167, 169, 197, 207
sinus, 107
skeletal
muscles, 60, 66,
system, 32, 37–54
skin, 33, 195–213, 221, 234
callus, 44, 65
cells, 196–200
color, 198–199
growth, 196
nerves, 197, 201
sensitivity, 206–207
temperature, 124, 202–203
thickness, 197
skull, 38, 43, 45–46, 169
sphincter
gastroesophageal, 74–75
ileocecal, 78
pyloric, 76, 157
spinal
column, 47, 167–168
cord, 61, 149–150, 152, 166–169, 234
nerves, 152, 167–168
stapes, 47, 182
stem cells, 129, 233, 239
sternum, 45, 48
stomach, 69, 74–76, 157
striations, 60, 66
subcutaneous tissue, 206
sutures, 43, 46
sweat, 80–81, 202
glands, 196–197, 201–202